2nd International Conference on Transport and Storage of Hydrocarbons 2018

IOP Conference Series: Materials Science and Engineering
Volume 445

Tyumen, Russia
29 – 31 August 2018

ISBN: 978-1-5108-8013-9
ISSN: 1757-8981

Printed from e-media with permission by:

Curran Associates, Inc.
57 Morehouse Lane
Red Hook, NY 12571

Some format issues inherent in the e-media version may also appear in this print version.

Printed by Curran Associates, Inc. (2019)

For permission requests, please contact the Institute of Physics at the address below.

Institute of Physics
Dirac House, Temple Back
Bristol BS1 6BE UK

Phone: 44 1 17 929 7481
Fax: 44 1 17 920 0979

techtracking@iop.org

Additional copies of this publication are available from:

Curran Associates, Inc.
57 Morehouse Lane
Red Hook, NY 12571 USA
Phone: 845-758-0400
Fax: 845-758-2633
Email: curran@proceedings.com
Web: www.proceedings.com

Table of contents

Volume 445

International Conference on Transport and Storage of Hydrocarbons

29–31 August 2018, Tyumen, Russian Federation

Accepted papers received: 31 October 2018

Prelims

Papers

Preface

The International Conference "Transport and Storage of Hydrocarbons" was held in Tyumen, Russia.

The 2018 Conference is aiming to convene undergraduates, post-graduates and researchers from Russia and other countries to present their research works and share their ideas on the most urgent topics in the sphere of transportation and storage of hydrocarbons.

The three-day Conference brought together the specialists from different countries and organizations and provided an excellent opportunity for knowledge exchange, making oral contributions and poster presentations, and initiating discussions on the topics that are of interest to the Conference participants.

The selected papers of the Conference cover a wide range of technical areas and modern aspects of transportation and storage of hydrocarbons.

Peer review statement

All papers published in this volume of *IOP Conference Series: Materials Science and Engineering* have been peer reviewed through processes administered by the proceedings Editors. Reviews were conducted by expert referees to the professional and scientific standards expected of a proceedings journal published by IOP Publishing.

Calculation of transverse vibrations of the pump set shaft

V M Loginovskikh, D A Cherentsov and K S Voronin

Industrial University of Tyumen, 38 Volodarskogo st., Tyumen, 625000, Russia

E-mail: girl_mail.ru94@mail.ru

Abstract. The paper presents calculations of the transverse vibrations of the shaft of the pumping unit, represented as a rod (beam), on the basis of the Bubnov-Galerkin method. An estimate is made of the convergence of the results obtained, from which the main conclusions are formulated.

1. Introduction

Reliability of pumps, like any machinery and mechanisms, is determined by several factors. First, it depends on the perfection of the design and, accordingly, the completeness of its adaptation to the operating conditions of the machines. Second, it is determined by the quality of factory manufacturing. And, finally, it depends on the operating conditions of the pumps.

Increasing the reliability of pumping units due to the consideration of the first two groups of factors finds its quite successful solution in engineering organizations - design bureaus and factories. However, the degree of success of these solutions largely depends on how well the pumps are able to adapt to the possible changes in the operating conditions occurring in the conditions of their operation quite often. Not only does the operational reliability of the pump units significantly depend on this, but also the economy of their operation.

Increased vibration is among the factors most adversely affecting the reliability of pumping units. In modern conditions of operation and management of transport, oil and gas storage systems, as well as elimination of accidents, significant importance is placed on the issues of management, organization and control of technological processes.

With the toughening of operating conditions, the intensity and frequency of vibrations of pumping units are increasing, and the pressure pulsations that lead to breakdown of pumping units and, consequently, breakdown in the technological process have a negative effect. In connection with which the requirements for the vibration protection of pumping units are sharply increasing. However, research in this area is increasingly reduced to determining the frequencies of free oscillations that do not depend on the characteristics of the shaft. Therefore, research in this area is relevant.

2. Methodology

When deriving the equation of transverse vibrations of a rod (or beam), we shall assume that in the undeformed state the so-called elastic axis of the rod is rectilinear and coincides with the line of centers of gravity of the cross sections of the rod. We take this rectilinear axis as the coordinate axis x: and from it we will count the deviations of the rod elements for transverse vibrations. We will assume, at least at first, that the deviations of the individual points of the axis of the rod occur perpendicular to the rectilinear, undeformed direction, neglecting the displacements of these points parallel to the axis.

Further, we assume that the deviations of the points of the axis of the rod in transverse vibrations occur in the same plane ("plane of oscillations") and are "small" deviations in the sense that the restoring forces that arise remain within the limits of proportionality.

Under such assumptions, the deviations of the points of the axis of the rod for transverse vibrations are uniquely determined by a single function of two variables, the coordinate x and the time t:

$$y = y(x,t) \tag{1}$$

This function satisfies a linear partial differential equation of the fourth order, which can be constructed as follows.

The equation that allows us to find transverse oscillations with the constant rigidity EJ

$$\rho\frac{\partial^2 w}{\partial t^2} + \frac{\partial^2}{\partial x^2}\left(\frac{EJ}{L^4}\frac{\partial^2 w}{\partial x^2}\right) = 0 \tag{2}$$

where ρ - density (kg/m^3), EJ - stiffness on the deflection [E (Pa) - modulus of elasticity, J (m^4) - moment of inertia of the cross section of the rod with respect to the central axis of the cross section perpendicular to the plane of oscillations], L - length of the shaft m).

The obtained differential equation of the beam oscillations contains the second time derivative. Therefore, for t = 0, it is necessary to put two initial conditions:

$$w(x,0) = w_0(x), \tag{3}$$

$$\dot{w}(x,0) = 0 \tag{4}$$

In the variable x, in the equation, there is the fourth-order derivative, therefore, at each end of the beam, in accordance with the conditions for its fixing, it is necessary to put two boundary conditions.

$$w(0,t) = 0,\ \dot{w}(0,t) = 0 \tag{5}$$

$$\ddot{w}(1,t) = 0,\ \ddot{w}(1,t) = 0 \tag{6}$$

The Bubnov-Galerkin method is used for the solution. The required displacements are specified as:

$$u = a_1(t)b_1(x) + a_2(t)b_2(x) + \cdots a_n(t)b_n(x) = \sum_{i=1}^{n} a_n(t)b_n(x) \tag{7}$$

where a_1, a_2 ... a_n are unknown functions of the variable t; b_1, b_2 ... b_n are given basis functions of the variable x.

We substitute u into the equations of motion. Since the displacements (7) are not an exact solution of the equations of motion, on the right-hand side, instead of zero, we obtain the discrepancy R_1:

$$\mu(x)\frac{\partial^2(a_n(t)b_n(x))}{\partial t^2} + EJ\frac{\partial^4(a_n(t)b_n(x))}{\partial x^4} = R_1 \tag{8}$$

In accordance with the Bubnov-Galerkin method, it is required that orthogonality conditions of the residual R_1 of the basis function b_n be satisfied.

A basis function satisfying the main boundary conditions

$$b_n(x) = \sin\frac{m\pi x}{L};\ n = 1, \dots m \tag{9}$$

The orthogonality condition can be written in the following form:

$$\mu_x\int_0^L\left(\sum b_n\ddot{a}_n\right)b_n dx + EJ\int_0^L\left(\sum a_n\frac{\partial^4 b_n}{\partial x^4}\right)b_n dx = 0 \tag{10}$$

$$\begin{cases} \mu_x \int_0^L \left(\sum b_n \ddot{a}_n\right) b_1 dx + EJ \int_0^L \left(\sum a_n \frac{\partial^4 b_n}{\partial x^4}\right) b_1 dx = 0 \\ \dots\dots\dots \\ \mu_x \int_0^L \left(\sum b_n \ddot{a}_n\right) b_m dx + EJ \int_0^L \left(\sum a_n \frac{\partial^4 b_n}{\partial x^4}\right) b_m dx = 0 \end{cases} \quad (11)$$

$$\begin{cases} \ddot{a}_1 \mu_x \int_0^L b_1 b_1 dx + a_1 EJ \int_0^L \frac{\partial^4 b_1}{\partial x^4} b_1 dx + \dots + \ddot{a}_n \mu_x \int_0^L b_n b_1 dx + \\ \quad + a_n EJ \int_0^L \frac{\partial^4 b_n}{\partial x^4} b_1 dx = 0 \\ \dots\dots\dots \\ \ddot{a}_1 \mu_x \int_0^L b_1 b_n dx + a_1 EJ \int_0^L \frac{\partial^4 b_n}{\partial x^4} b_1 dx + \dots + \ddot{a}_n \mu_x \int_0^L b_n b_m dx + \\ \quad + a_n EJ \int_0^L \frac{\partial^4 b_n}{\partial x^4} b_m dx = 0 \end{cases} \quad (12)$$

Equations are a system of m ordinary linear differential equations with constant second-order coefficients with respect to unknown functions $b_1(t)$, ... $b_m(t)$.

The condition for the existence of a nontrivial solution is that the principal determinant of the system disappears.

To solve the proposed method, it is necessary to determine the minimum number of basis functions that will allow calculating parameters without loss of accuracy. Let us estimate how the values of the parameters change when the number of basis functions changes (from 1 to 10).

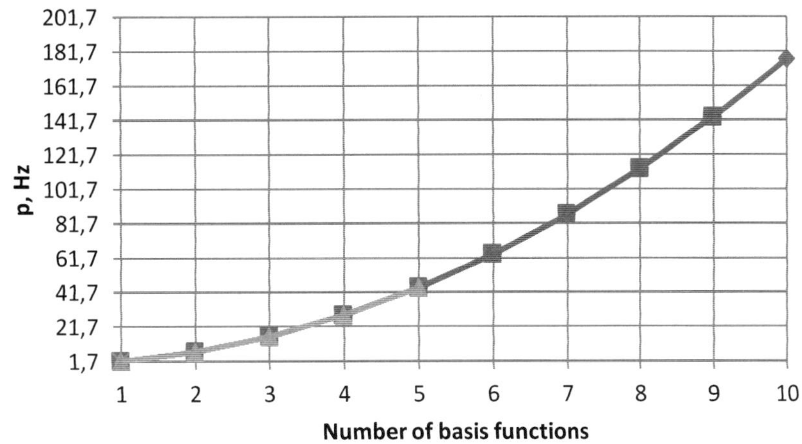

Figure 1. Evaluation of the convergence of the oscillation frequency.

3. Conclusion

Analyzing the obtained values, it can be concluded that when the number of basis functions increase, the values of oscillation parameters tend to a certain limit. Deviation between decisions in which seven and eight functions were considered is 0.02%. Deviation at four and five terms is 0.01%. To obtain satisfactory results, it is sufficient to retain two functions.

References

[1] Andrievsky B R and Fradkov A L 2001 *Elements of Mathematical Modeling in Software Environments MATLAB 5 and Scilab* (St. Petersburg: Nauka) 287 p
[2] Wolfson I I 2017 *Dynamics of Machines. Fluctuations* 2nd ed (Moscow: Urait) 275 p
[3] Gumerov A G, Gumerov R S and Akberdin A M 2003 *Diagnostics of Equipment of Oil Pumping Stations* (Moscow: Nedra) 348 p
[4] Zakharenko S O 2007 *Coll. Materials of the All-Russian Sc. and Pr. Conf. Actual Problems of Construction, Ecology and Energy Saving in Western Siberia* (Tyumen: ColeSo)
[5] Iglin S P 2005 *Mathematical Calculations on the Basis of Matlab* (St. Petersburg: BHV-Peterburg) 634 p
[6] Levin V E 2009 (Novosibirsk: Novosibirsk State Technical University) 138 p
[7] Lobodenko E I 2011 *Oscillations and Vibrodiagnostics of Main Pumping Units* Part 2 (Tyumen: TyumGASU) 34 p
[8] Panovko Ya G 2010 *Fundamentals of the Applied Theory of Oscillations and Shocks* (Moscow: Librocom) 274 p
[9] Khismatullin Sh Sh 2015 *Protection from Vibration in Industries and Construction* (Orenburg: Orenburg State University, EBS ASV) 291 p
[10] Shampine L M, Gladwell I and Thompson S 2009 *Solving ODEs with Matlab* (St. Petersburg: Lan) 299 p
[11] Yablonskiy A A and Noreyko S S 2007 *Course of the Theory of Oscillations* 5th ed (St. Petersburg: BHV-Peterburg) 336 p

Estimating the cost of constructing and operating a section of a pipeline in the search for its optimal route

K S Voronin [1, a], P V Grigorieva [1, b] and D A Cherentsov [1, c]

[1] Industrial University of Tyumen, 38 Volodarskogo str., Tyumen, 625000, Russia

E-mail: [a]voronin_tsogu@mail.ru, [b]polino4ka.g@yandex.ru, [c]cherentsovda@bk.ru

Abstract. The paper considers the calculation of the relative cost of construction and operation of a pipeline section in the search for its optimal route. Comparison of efficiency of application of various options of data representation on unit cost of pipeline route depending on topographical features.

1. Introduction

When designing pipelines, engineers face the task of finding the optimal route, ensuring minimum costs for their construction and further operation [1, 4]. This task is very time consuming, as it requires consideration of many feasible options for passing the route and taking into account a large number of factors affecting the cost of pipeline construction and operation. Currently, the choice of the route is most often based on the expert opinion of the designer.

In [6] an algorithm of searching for the optimum pipeline route in view of topographical features of of the area is offered. This algorithm allows one to significantly reduce the time input for design by automating the process of searching for the optimal route, and also to reduce the influence of the human factor. For effective operation of this algorithm, it is required to estimate the cost or weight of the considered pipeline route option most accurately.

2. Methodology

The weight of the line (pipeline section) can be defined as the sum of the weights of all infinitely small segments, of which this section consists. To do this, we integrate the specific weight of each point on the map along the length of the given line:

$$C = \int_0^L w dL \tag{1}$$

where C – total weight of the line in question;
L – length of the line in question;
w – specific weight of a point on a weight map;
dL – length of an elementary segment.

Since, as a result of the work of the route search algorithm, we get a broken line consisting of rectilinear segments, the total weight of this line can be determined as the sum of weights of these segments:

$$C_i = \int_{L_{i-1}}^{L_i} w dL \tag{2}$$

$$C = \sum_{i=1}^{n} C_i = \sum_{i=1}^{n} \int_{L_{i-1}}^{L_i} w dL \tag{3}$$

where C_i – weight of the ith segment;
L_i – length i of the first segments;
n – number of segments.

To determine the cost of the route of the designed pipeline section, it is necessary to draw up a weight map showing the unit cost (specific weight) of the route passing through a particular section of the terrain. The weight map can be presented in vector or raster versions. Each of these options has its own peculiarities for the calculation.

The weight map, represented in a vector form (Figure 1), is divided into several closed areas, for each of which the specific cost of laying the pipeline (specific weight) is indicated.

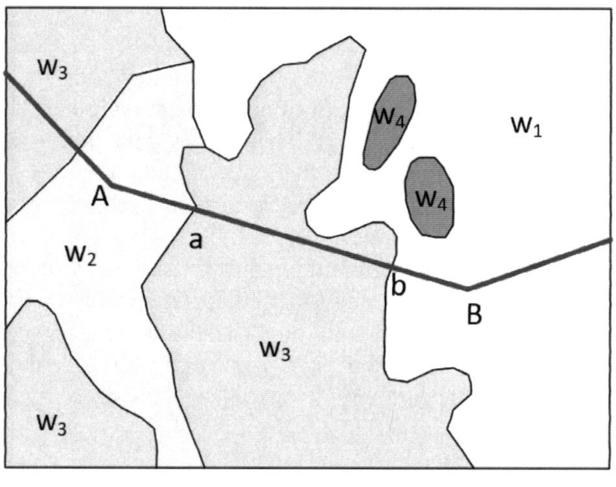

Figure 1. Vector map of weights.

In this case, the weight of the pipeline section is determined by summing the products of the weights of the map areas by the lengths of the segments located in these areas:

$$C = \sum_{i=1}^{n} w_i L_i = \sum_{i=1}^{n} w_i \sqrt{(x_i - x_{i-1})^2 + (y_i - y_{i-1})^2} \tag{4}$$

where w_i – specific weight of the area in which the ith segment is located;
L_i – length of the ith segment;
x_i, y_i – coordinates of the ith point.

A raster map of weights (Figure 2) is a table for each cell of which the value of the specific cost of pipeline laying (specific weight) is indicated.

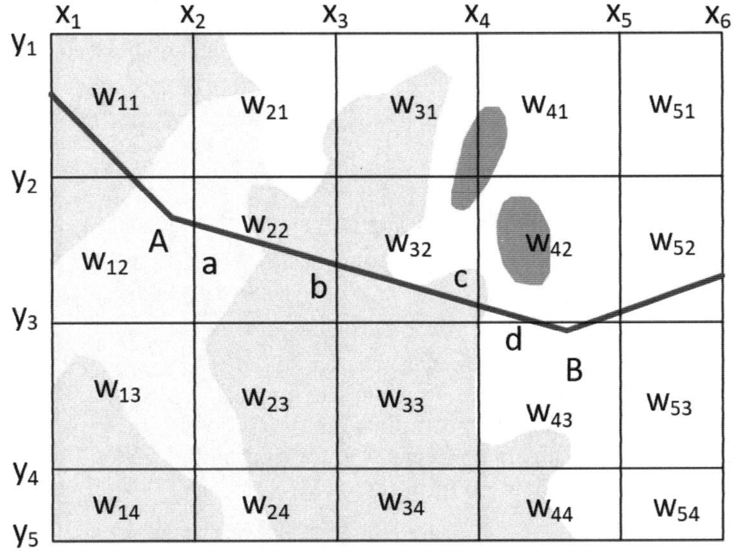

Figure 2. Raster map of weights with cells.

The weight of the pipeline section, as in the previous case, can be determined by the formula (4), adding up the products of the specific cell weights by the lengths of the segments located inside these cells. However, in comparison with the calculation of a vector map of weights, this variant requires considerably less processing power to determine the coordinates of the intersection points of the line with the boundaries of the cells of the map.

To ensure that there are no abrupt changes in the specific weight at the borders of the cells of the weight map, it can be specified not for the entire cell, but only for nodes, that is, the intersection points of the grid lines (Figure 3a). In this case, the specific weight of any intermediate point can be found by means of bilinear interpolation (Figure 3b).

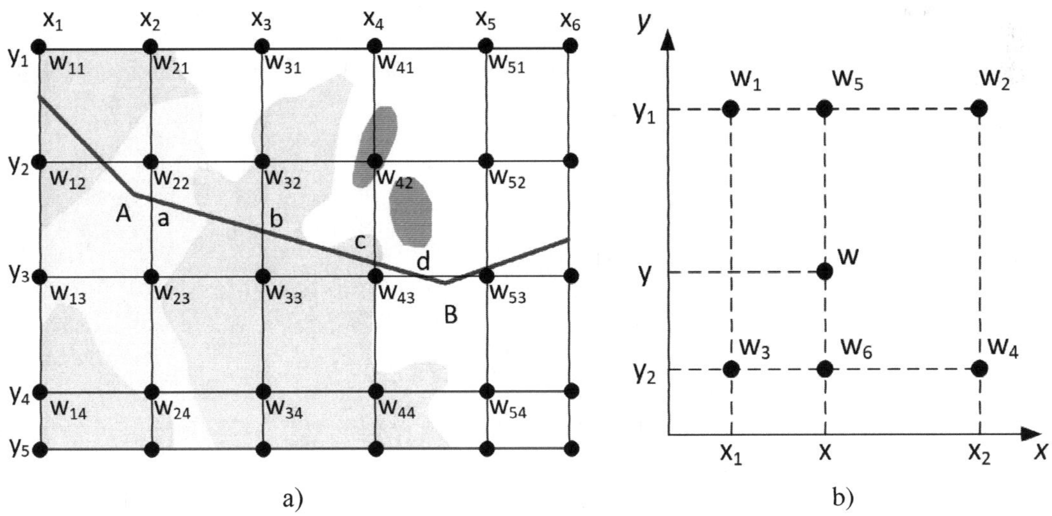

Figure 3. Raster map of weights with nods.
a) general view of the map;
b) finding the specific weight at the intermediate point.

Let us consider one grid cell. The specific weight w at a point with coordinates (x, y) can be determined by knowing the weights of the node points w_1, w_2, w_3, w_4 and their coordinates:

$$w = w_5 + (w_6 - w_5)\frac{y - y_1}{y_2 - y_1} \tag{5}$$

$$w_5 = w_1 + (w_2 - w_1)\frac{x - x_1}{x_2 - x_1} \tag{6}$$

$$w_6 = w_3 + (w_4 - w_3)\frac{x - x_1}{x_2 - x_1} \tag{7}$$

where $w_1, \ldots, w_6$ – specific weights at points 1 to 6;
x_1, x_2, y_1, y_2 – cell boundary coordinates.

Substituting (6), (7) in (5), we obtain an expression for finding the specific weight at an arbitrary point inside the cell under consideration:

$$w = w_1 + (w_2 - w_1)\frac{x - x_1}{x_2 - x_1} + \left(w_3 - w_1 + (w_4 - w_3 - w_2 + w_1)\frac{x - x_1}{x_2 - x_1}\right)\frac{y - y_1}{y_2 - y_1} \tag{8}$$

We substitute the obtained expression for the specific weight (8) into the initial equation to determine the weight of the segment (2) and transform:

$$C = \int \left(w_1 + (w_2 - w_1)\frac{x - x_1}{x_2 - x_1} + \left(w_3 - w_1 + (w_4 - w_3 - w_2 + w_1)\frac{x - x_1}{x_2 - x_1}\right)\frac{y - y_1}{y_2 - y_1}\right) dL$$

$$C = w_1 \int dL + \frac{w_2 - w_1}{x_2 - x_1}\int (x - x_1)dL + \frac{w_3 - w_1}{y_2 - y_1}\int (y - y_1)dL + \\ + \frac{w_4 - w_3 - w_2 + w_1}{(x_2 - x_1)(y_2 - y_1)}\int (x - x_1)(y - y_1)dL \tag{9}$$

We decompose the elementary segment dL into horizontal dx and vertical dy components (Figure 4).

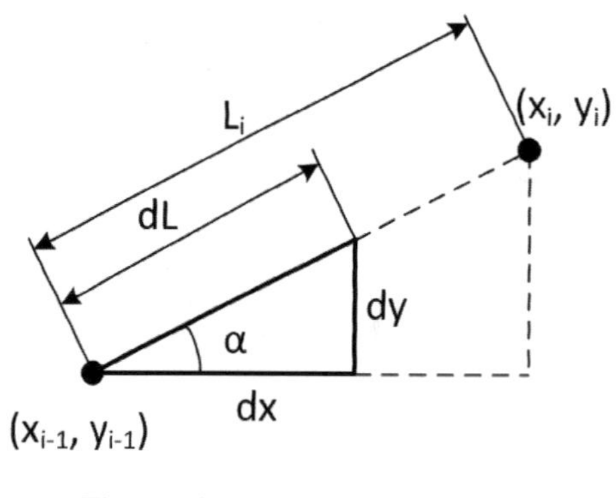

Figure 4. Elementary segment.

Let us express the length of an elementary segment through its projections onto the horizontal and vertical axes:

$$dL = \frac{dx}{\cos\alpha} = \frac{dy}{\sin\alpha}$$

$$\cos\alpha = \frac{(x_i - x_{i-1})}{\sqrt{(x_i - x_{i-1})^2 + (y_i - y_{i-1})^2}}; \qquad \sin\alpha = \frac{(y_i - y_{i-1})}{\sqrt{(x_i - x_{i-1})^2 + (y_i - y_{i-1})^2}}$$

$$dL = \frac{\sqrt{(x_i - x_{i-1})^2 + (y_i - y_{i-1})^2}}{(x_i - x_{i-1})} dx = \frac{\sqrt{(x_i - x_{i-1})^2 + (y_i - y_{i-1})^2}}{(y_i - y_{i-1})} dy \qquad (10)$$

We write the equation of the straight line in a general form and substitute the coordinates of the beginning and end of the segment:

$$y = ax + b$$

$$\begin{cases} y_{i-1} = ax_{i-1} + b \\ y_i = ax_i + b \end{cases}$$

$$a = \frac{y_i - y_{i-1}}{x_i - x_{i-1}}; \qquad b = y_i - \frac{y_i - y_{i-1}}{x_i - x_{i-1}} x_i$$

$$\begin{cases} y = \dfrac{y_i - y_{i-1}}{x_i - x_{i-1}} x + y_i - \dfrac{y_i - y_{i-1}}{x_i - x_{i-1}} x_i \quad , \quad if\ x_{i-1} \neq x_i \\ x = x_i \quad , \quad if\ x_{i-1} = x_i \end{cases} \qquad (11)$$

Substituting expressions (10), (11) in (9), we can obtain a formula for determining the weight of a segment inside one cell of the map. If the segment is not vertical, that is, $x_{i-1} \neq x_i$, then:

$$C = \frac{\sqrt{(x_i - x_{i-1})^2 + (y_i - y_{i-1})^2}}{(x_i - x_{i-1})} \left(w_1 \int dx + \frac{w_2 - w_1}{x_2 - x_1} \int (x - x_1) dx + \frac{w_3 - w_1}{y_2 - y_1} \int (ax + b - y_1) dx + \frac{w_4 - w_3 - w_2 + w_1}{(x_2 - x_1)(y_2 - y_1)} \int (x - x_1)(ax + b - y_1) dx \right)$$

$$C_i = \frac{\sqrt{(x_i - x_{i-1})^2 + (y_i - y_{i-1})^2}}{(x_i - x_{i-1})} \left(w_1 \int_{x_{i-1}}^{x_i} dx + \frac{w_2 - w_1}{x_2 - x_1} \int_{x_{i-1}}^{x_i} (x - x_1) dx + \frac{w_3 - w_1}{y_2 - y_1} \int_{x_{i-1}}^{x_i} (ax + b - y_1) dx + \frac{w_4 - w_3 - w_2 + w_1}{(x_2 - x_1)(y_2 - y_1)} \int_{x_{i-1}}^{x_i} (ax^2 + (b - y_1 - ax_1)x + (y_1 - b)x_1) dx \right) \qquad (12)$$

After integrating, we obtain:

$$C_i = \sqrt{(x_i - x_{i-1})^2 + (y_i - y_{i-1})^2}\left(w_1 + \frac{w_2 - w_1}{x_2 - x_1}\left(\frac{x_i + x_{i-1}}{2} - x_1\right) + \frac{w_3 - w_1}{y_2 - y_1}\left(\frac{a(x_i + x_{i-1})}{2} - (b - y_1)\right) + \frac{w_4 - w_3 - w_2 + w_1}{(x_2 - x_1)(y_2 - y_1)}\left(\frac{a\left(x_i^2 + x_i x_{i-1} + x_{i-1}^2\right)}{3} + \frac{(b - y_1 - a x_1)(x_i + x_{i-1})}{2} + (y_1 - b)x_1\right)\right) \tag{13}$$

If the segment is vertical, that is $x_{i-1} = x_i$, then:

$$C = \frac{\sqrt{(x_i - x_{i-1})^2 + (y_i - y_{i-1})^2}}{(y_i - y_{i-1})}\int\left(w_1 + (w_2 - w_1)\frac{x - x_1}{x_2 - x_1} + \left(w_3 - w_1 + (w_4 - w_3 - w_2 + w_1)\frac{x - x_1}{x_2 - x_1}\right)\frac{y - y_1}{y_2 - y_1}\right)dy$$

$$C_i = \frac{\sqrt{(x_i - x_{i-1})^2 + (y_i - y_{i-1})^2}}{(y_i - y_{i-1})}\left(\left(w_1 + (w_2 - w_1)\frac{x_i - x_1}{x_2 - x_1}\right)\int_{y_{i-1}}^{y_i} dy + \frac{\left(w_3 - w_1 + (w_4 - w_3 - w_2 + w_1)\frac{x_i - x_1}{x_2 - x_1}\right)}{y_2 - y_1}\int_{y_{i-1}}^{y_i}(y - y_1)dy\right) \tag{14}$$

After integrating, we obtain:

$$C_i = \sqrt{(x_i - x_{i-1})^2 + (y_i - y_{i-1})^2}\left(w_1 + (w_2 - w_1)\frac{x_i - x_1}{x_2 - x_1}\right) + \frac{\left(w_3 - w_1 + (w_4 - w_3 - w_2 + w_1)\frac{x_i - x_1}{x_2 - x_1}\right)}{y_2 - y_1}\left(\frac{y_i + y_{i-1}}{2} - y_1\right) \tag{15}$$

Formulas (13), (15) allow calculating the weight of a pipeline section using a variant of a map of weights with smooth transitions between nodes. However, this option requires more processing power than the option with the weight of the cells of the map.

3. Conclusion

Thus, the use of a vector map of weights allows the most accurate estimate of the cost of the route of the pipeline section, but because of the difficulty in determining the coordinates of the intersection of the segments of the route with the boundaries of the map areas, it requires more computational resources and time for making calculations. Moreover, the time spent directly depends on the area of the considered section of the map (on the number and complexity of the selected areas).

The use of raster variants of the weight map significantly reduces the time spent on calculations, but gives a less accurate result due to the error in determining the boundaries of map areas with different specific weights. When the resolution of the map increases, the error decreases, but the memory input and the calculation time of the pipeline route cost increase.

References

[1] Babin L A 1970 *Selection of Main Pipeline Routes* (Leningrad: Nedra) 128 p
[2] Belousov A I and Tkachev S B 2006 *Discrete Mathematics* (Moscow: MSTU) 744 p
[3] Borodavkin P P and Berezin V 1987 *Construction of Main Pipelines* 2nd ed (Moscow: Nedra) 471 p
[4] Borodavkin P P, Berezin V L and Ruderman S Yu 1974 *Selection of the Optimal Routes of Main Pipelines* (Moscow: Nedra) 240 p
[5] Dubrovin D D and Zemenkova M Yu 2017 Analysis of the problems of ensuring minimum distances from the objects of main gas pipelines *Oil and Gas Terminal, Coll. of Sc. Works of the Intern. Sc. and Techn. Conf. of Grad. Students n. a. Prof. N A Malyushin* pp 72-7
[6] Grigorieva P V, Voronin K S and Cherentsov D A 2017 Selection of the pipeline route, considering the peculiarities of the terrain *Oil and Gas Terminal, Coll. of Sc. Works of the Intern. Sc. and Techn. Conf. of Grad. Students n. a. Prof. N A Malyushin* pp 56-60.
[7] Mutavaliev Sh T, Zemenkov Yu D and Zemenkova M Yu 2015 Construction, design of pipelines in the mountainous terrain and their optimization tasks *Oil and Gas Terminal, Coll. of Sc. Works of the Intern. Sc. and Techn. Conf.* ed S Yu Podorozhnikov pp 132-5
[8] Smirnov A N, Podorozhnikov S Yu, Zemenkov Yu D and Shipovalov A N 2011 Increase of reliability and efficiency of the technological regime of the oil collection network *Pipeline Transport: Theory and Practice* 1 (23) pp 27-9
[9] Spasskaya K A 2014 Criteria for choosing a variant of the direction of the railway route, considering the development of landslide processes *Internet J. Naukovedenie* 3
[10] Zemenkova M Y, Zemenkov Y D фтв Shipovalov A N 2016 Mathematic modeling of complex hydraulic machinery systems when evaluating reliability using graph theory *IOP Conf. Ser. Mat. Sc.and Eng. International Scientific and Practical Conference on Urgent Problems of Modern Mechanical Engineering* 012056.

Oil and gas pipeline system of the Tyumen Oblast: social and spatial analysis

M G Ganopolsky[2] and L M Markova[1]

[1]Industrial University of Tyumen, 38 Volodarskogo str., Tyumen, 625000, Russia
[2]Institute of the problems of Northern development, Tyumen Scientific Centre SB RAS, Russia

E-mail: gmichaelg@mail.ru; markova@tyuiu.ru

Abstract. The article describes the spatial analysis of the trunk oil and gas pipeline system running on the territory of the Tyumen Oblast, in combination with the near-route populated utilitarian settlements forming a unique territorial production cluster.

1. Introduction

Pipelines that originate in the Tyumen oil and gas fields represent a unique transport system both in terms of their production capabilities and in terms of the resonant impact on the character of the region's economic and social development [1], [2]. Throughout their entirety within the boundaries of the Tyumen Oblast, they have become an integral feature of its industrial landscape, an important factor in the organization of social space. At the same time, it is the spatial aspect of the functioning of this system that remains outside the field of view of researchers. It would seem that this aspect is closest to the problem of transport geography. But here the main works (not so numerous) are concentrated on two main areas: the interaction of transport and settlement [3] and the morphology of transport networks [4]. And both these areas are connected, as a rule, with more branched networks of automobile and railway transport. It is clear that the tools used in this case are not suited to studying the peculiarities of pipeline transport. Some efforts are required to modify and adapt the methods and models used. Such an attempt is made in our work.

2. Methods

In this case, the concept of the support frame put forward in his time by N.N. Baran and developed by his disciples and followers may prove promising [5]. It allows one to combine the advantages of both morphological and settlement areas in the geography of transport and at the same time consider the region as an integral structural and communicative formation.

An analysis of the structure of the settlement of the modern Tyumen region shows that along with the main frame, which unites large oil and gas production centers and support-rear facilities in the south of the region, there are also relatively autonomous settlement clusters. One of them is represented by the settlements that gravitate towards the main oil and gas transportation network. And although in the economic-geographical in administrative-territorial, communicative and other relations these settlements are sometimes presented as unconnected, the commonality of their production purpose and socio-cultural appearance is unquestionable. Nevertheless, this cluster is an almost unexplored socio-spatial phenomenon, and we see our task in proposing an initial theoretical sketch of this phenomenon and thereby set the impulse for its further research.

The autonomy of the cluster under consideration is due not only to the specifics of the territorial organization of the industry subsystems that make up the oil and gas complex, but also to the peculiarities of the settlement of these areas. The urbanized settlement system was formed in the Tyumen region in just three to four decades in the course of a new industrial development in a sparsely populated, and in some places uninhabited space. Unlike the old-developed regions, where such systems mature not so rapidly and, as a rule, on the basis of rural settlement, the picture of urbanization in the Tyumen region was different. The pattern of settlement here was first drawn up sketchily, then projected onto a supposedly clean space, more consistent with the matrix of geological markup of the territory than with the previously established settlement scheme. This was due to the involvement in the economic turnover of an extensive oil and gas province, with its intensive industrial development, based on the location of production facilities confined to hydrocarbon deposits. People were seen as resources mobilized and distributed in the necessary amount for their erection and functioning. As a result, the formed system of settlement and its support framework acquired characteristic features of utilitarianism [6].

As for the main transport of oil and gas, the basic principle of its development at the time of the beginning of the development of oil and gas fields was based on the logic of centralized management of the national economy and the existing specialization of the economic regions of the country. It was believed that the enterprises for the preparation and processing of hydrocarbon raw materials should be concentrated in places of mass consumption, remote from the mining areas by thousands of kilometers. In addition, the essential in this strategy was the export of oil and gas outside the country. An accelerated construction of ultra-long oil and gas pipelines was required, which to this day determine the configuration of the main pipeline network in Russia and CIS countries.

However, from the point of view of the organizing influence on the regional settlement system, the ground objects of the transport network are most significant. These are, first of all, oil pumping stations and gas compressor stations (OPSs and CSs). There are a number of methods for arranging the OPSs and CSs on the pipeline route [7], [8]. However, theoretical calculations is only a preliminary outline of their location. The practice of building pipelines is based on the fact that pumping stations are preferable to locate near settlements, sources of energy and water supply, the existing network of railways and highways. In addition, certain requirements are also imposed on the sites of these stations. Thus, in the developed regions their location is set initially. It is clear that everything was more complicated in the Tyumen Oblast. Since the upper limit of the normative distance between two neighboring stations can not be exceeded, the choice of their locations was often associated with the need to build new settlements, build a minimum of production and social infrastructure facilities, and redo the majority of communications. But even in the existing villages, sometimes serious modernization of buildings, facilities, production facilities, and the whole life-support complex was required.

The main characteristics of pipeline transport in the Tyumen Oblast at the final stage of network formation are given in Tables 1 and 2 [9].

Table 1. Oil pipelines of the Tyumen Oblast.

Main characteristics	1980	1985	1990	1995
Length of the route, *km*	3395	4204	6178	6085
In a single-line version, *km*	6275	8159	10684	10597
Number of oil pipelines, *pcs*	12	20	32	34
Average diameter of pipes, *mm*	1057	1029	988	988
Number of OPSs, *pcs*	60	80	83	83

Table 2. Gas pipelines of the Tyumen Oblast.

Main characteristics	1980	1985	1990	1995
Length of the linear part, *km*	9279	18024	27672	27790
Number of gas pipelines, *pcs*	7	12	18	20
Average diameter, *mm*	1320	1369	1386	1395
Number of CSs, *pcs*	27	42	50	50

Since then, the number of settlements with ground network facilities that require constant stationary services has remained virtually unchanged. But the ratio of the generators of settlement factors has changed. However, this almost did not affect cities, but in many villages and towns, including suburban ones, the factor of servicing the transport main became predominant, and sometimes the only factor.

It would seem that these are weighty arguments in favor of considering a network of pipelines and near-route settlements as a linear part and nodal elements of a supporting framework for settlement. But, despite the obvious organizing importance of pipelines, this structure does not form the framework of settlement in its classical sense. Oil and gas mains can not be called communication routes in the usual sense, as well as technological roads along the pipeline route. Passenger and cargo communication between rural settlements is organized via other roads, and in some cases carried out by air transport. Therefore, the nodal elements of such a framework are deprived of two basic functions: region-forming and communicative (interaction factor). That is, they have only a development role [4]. This is mainly the expansion of the zone of influence of the settlement on the adjacent territory, its involvement in economic circulation, as well as the use for other purposes, including recreational ones. In this regard, the entire structure of this cluster by its characteristics is closest to the development frame. But this is not its only purpose. This structure is unique in that it combines both generative and conservative mechanisms of impact on the population. This is a kind of a proto-frame. On the one hand, it facilitates the expansion of the area of territory settlement, the emergence of new accommodation microcracks within the administrative district and at the inter-district level, and, on the other hand, preserves and consolidates the dominant position of the near-route settlements in the expanding settlement scheme.

The question of the place and role of this cluster in the socio-cultural space of the region deserves special consideration. An analogy with closed cities, or, more precisely, with settlements that fall under the category of closed administrative-territorial entities (CATE), may be useful here. Earlier, these cities and towns with a strict regime of secrecy were known only from the rumors. Now they are much more open and even became the subject of special studies (including geographic ones). Their closed nature is officially stipulated and provides for a number of prohibitions and restrictions.

It is clear that in small settlements along the pipeline route, isolation and closed nature are of a completely different kind. This, in fact, is "another Tyumen region," not so populous (here live less than 10% of the population of the region), little-known, unattractive, deprived of public attention. But it deserves such attention. Like closed cities, this place of life and work is specialized, it closely intertwines everyday life and a high degree of responsibility for the results of work. A lot depends on the social well-being of these places, including the energy and environmental security of not only the specific region, but the whole country [1], [10].

3 Conclusions

The cluster considered in this paper has a number of features. This combination of unique (single) and universal (typical) factors for a whole range of parameters: severe natural and climatic conditions, the scale of construction and operation of facilities, utilitarian structure of settlement and reproducing its social infrastructure, etc. All this turns it into a kind of representative facility, a kind of model of theoretical analysis and decision-making.

References

[1] Eremenko P T and Vorobyov N A 1989 *Development of Pipeline Transport in the USSR and Abroad* (Moscow: Nedra) 166 p
[2] Malyushin N A and Chepurskiy V N 1996 *Main Oil Pipelines of Western Siberia. Technical Condition. Reliability. Ecological Safety* (Tyumen: Pulse)
[3] Golts G A 1981 *Transport and Settlement* (Moscow: Nauka)
[4] Tarkhov S A 2005 *Evolutionary Morphology of Transport Networks* (Smolensk-Moscow: Universum)
[5] Lappo G M 1983 The concept of a supporting frame of the territorial structure of the national economy: development, theoretical and practical significance *Proc. of the USSR Acad. of Sc.. Geographic Ser.* 5 pp 16-28
[6] Ganopolskiy M G and Litenkova S P 2005 The structure of settlement in the Tyumen region: features of genesis and development prospects *Proc. of RAS. Geographic Ser.* 3 pp 56-62
[7] Aliev R A, Belousov V D and Nemudrov A G 1988 *Pipeline Transportation of Oil and Gas* (Moscow: Nedra)
[8] Tetelmin V V and Yazev V A 2013 *Main Oil and Gas Pipelines* (Moscow: Intellect) 352 p
[9] 2010 Operation of Main Oil Pipelines ed Yu D Zemenkov (Tyumen: TyumGNGU) 534 p
[10] Kuzmin E L and Kagramakov A K 2009 *Global Energy Security and Pipeline Transport. Political and legal aspect* (Voronezh: Science Book Publishing House) 256 p

Mathematic simulation of the effect of a buried oil pipeline on permafrost soils

V V Golik[1], B V Moiseev[1], S G Gulkova[2] and Yu D Zemenkov[1]

[1]Industrial University of Tyumen, 38 Volodarskogo st., Tyumen, 625000, Russia
[2]Far Eastern Federal University, 8 Sukhanova st., Vladivostok, 690950, Russia

E-mail: darkpaint@mail.ru, muzemenkova@mail.ru

Abstract. The paper is devoted to the study of the features of design and operation of main oil pipelines which run in the conditions of the Far North and places of permafrost soils distribution. The aim of the paper is developing of a mathematical model for the interaction of an underground oil pipeline in a multilayered concrete coated shell with permafrost soils, as well as a numerical solution of this model.

1. Introduction

At present, the geography of hydrocarbon production is changing, new fields of Eastern Siberia, the Far East, the Yamal Peninsula, the continental shelf of the Arctic seas are being developed according to the order of Government of Russia dated 13.11.2009 No. 1715-r "On the Energy Strategy of Russia for the period until 2030". The implementation of the strategy requires the design and construction of modern trunk oil pipelines for the transportation of hydrocarbons from production sites to domestic consumers under extremely unfavorable conditions [1].

Most of the above-mentioned territories refer to the places of permafrost soils distribution where the construction and operation of oil pipelines is hampered by unfavorable geo-cryological processes which can lead to a change in the design position of the pipeline and affect its reliability characteristics [2]. Therefore, even at the design stage, it is necessary to provide a set of protective measures to limit the impact of hazards within the limits which ensure reliable pipeline operation.

2. Methodology

The choice of technical measures to preserve the frozen condition of soils, preserve their temperature regime and ensure the reliability of the pipeline construction is made by calculation [3-6].

Data available for carrying out calculations are:

- climatic characteristics of the area;
- composition and temperature regime of the soil;
- technical characteristics of the pipe;
- physical and chemical parameters of the pumped product.

Under underground laying of oil pipelines, one of the main heat engineering calculations is the determination of heat losses. It causes a drop in product temperature, and as a result, a decrease in transport characteristics. The thermal losses at each section of the pipeline are determined by the formula:

$$q = \pi k\,(t_v - t_n), \tag{1}$$

where k – coefficient of the heat transfer from oil to the soil, t_v – temperature of the pumped medium, t_n – temperature of the surrounding soil.

Heat transfer coefficient k shows the amount of heat passing from the oil product to the frozen soil through the pipe wall and layers of insulation coating and is calculated by the formula:

$$k = \frac{1}{R}, \tag{2}$$

where R – thermal resistance of an isolated oil pipeline.

Thermal resistance is calculated for each layer of pipeline insulation coating:

$$R = \frac{1}{\alpha_1 D_v} + \frac{1}{2\lambda_{in}} \ln\frac{D_n}{D_v} \ldots + \frac{1}{\alpha_2 D_n}, \tag{3}$$

where D_n, D_v – the outer and inner diameters of the insulation layer, m, respectively, λ_{in} - thermal conductivity of the insulating layer at an average insulation temperature W/m°C, α_1 –coefficient of the heat transfer from the oil to the pipe wall W/m^2°C, α_2- coefficient of the heat transfer from the surface of insulation to the soil W/m^2°C.

Let us find the temperature of the surface of the wall of the underground oil pipeline:

$$T_{ow} = t_n + \frac{q}{\alpha_2 \pi D_n}\ [K], \tag{4}$$

Based on the results obtained, when calculating the temperature on the final layer of insulation (Figure 1), it is necessary to simulate the effect of the heat flow on the frozen soil surrounding the pipeline.

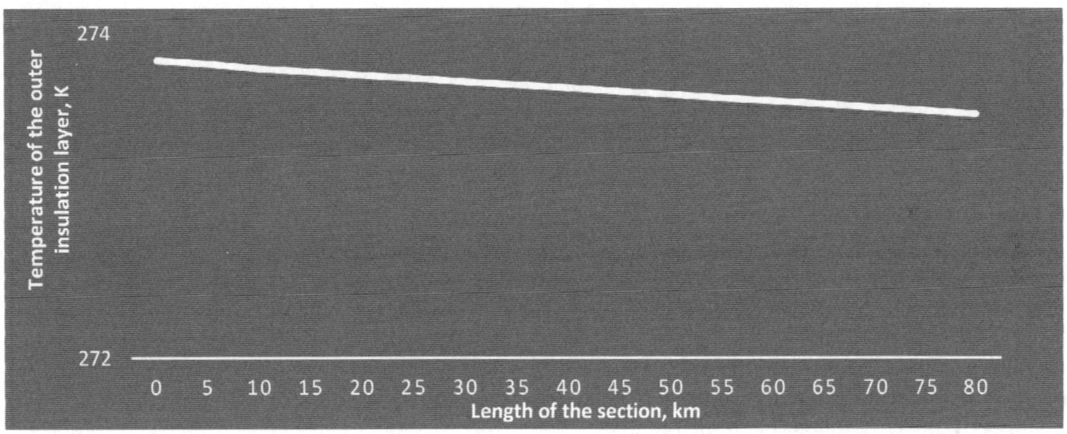

Figure 1. Change in the wall temperature along the length of the pipeline.

To obtain the parameters of the thawing aureole in the permafrost, a mathematical model of the thermal interaction of the pipeline with a radius r, buried to the depth H in the soil and describing there the heat exchange [7]. The choice of this model is due to the fact that the solution of the heat transfer problem the in the soil in a two-dimensional formulation will increase the velocity of the predicted calculations [8].

Let us consider a multilayered monolithic oil pipeline laid in permafrost with soil temperature t = -4°C at the laying depth for the Nadym Region of the Yamal-Nenets Autonomous District.

To set and solve the problem, the following assumptions were made: the average depth of the pipeline in the soil from 1 to 1.8 m; the soil was presented in the form of a square section of the frozen area around the pipeline, considering the effect of outside air on the soil; the outside air temperature was constant and equal to -20°C [9]. Considering the above-described conditions, the design model of the process of heat distribution around the underground oil pipeline is presented in Figure 2.

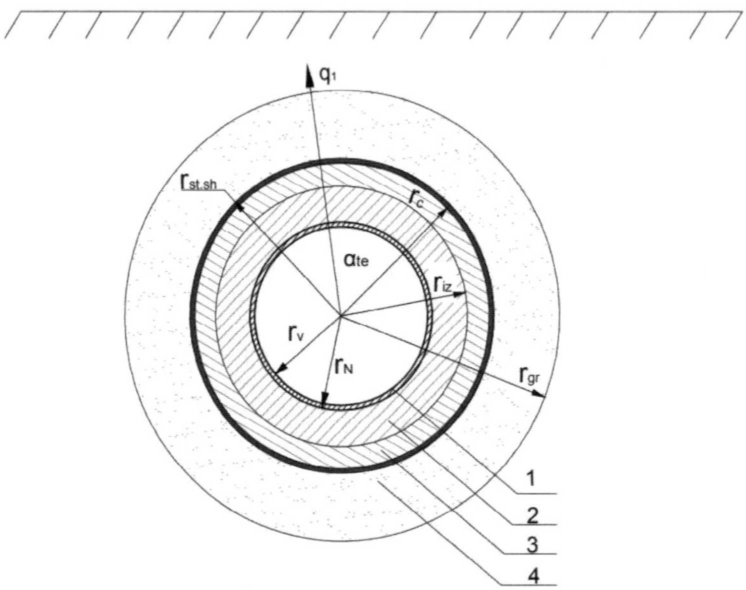

Figure 2. Calculation model of the thermal interaction of an underground oil pipeline in permafrost soils: 1. steel pipe; 2. thermal insulation of the foamed polyurethane; 3. nano-modified concrete cover; 4 permafrost soil.

$$C_{ef}\frac{\partial t}{\partial \tau} = \frac{\partial}{\partial x}\left(\lambda\frac{\partial t}{\partial x}\right) + \frac{\partial}{\partial y}\left(\lambda\frac{\partial t}{\partial y}\right), \quad (5)$$

where C_{ef} – effective heat capacity; t – soil temperature; τ – estimated time.

Also, the effective heat capacity and the temperature range Δt is determined at which the ice-water phase transformations occur in the soil:

$$C_{ef} = \begin{cases} c_{th}, & \text{when } t > t_f + \Delta t; \\ \rho c - \chi\gamma\frac{dW(t)}{d\tau}, & \text{when } t \in [t_f - \Delta t,\ t_f + \Delta t]; \\ c_{fr}, & \text{when } t > t_f - \Delta t. \end{cases} \quad (6)$$

Where ρ- soil density; t_f – temperature of moisture transition; χ- latent heat of phase transition; W – content of the non-frozen moisture; γ- specific gravity of the soil skeleton. The indices th and fr correspond to the thawing and frozen soil state:

$$\lambda = \begin{cases} \lambda_{th}, \text{when } t > t_f, \\ \lambda_{fr}, \text{when } t \le t_f. \end{cases} \quad (7)$$

Despite a fairly extensive set of precise analytical methods, many problems cannot be solved using them. Such problems are solved numerically. To solve the formulated model, the finite difference method was chosen. The calculation was performed in Excel and MATLAB & Simulink programs with subsequent verification (Figure 3).

Temperature conductivity	0,000645	Calc. for period	31540000 sec	8761,111 hour
Time increment	10			
Coordinate increment	1	Calculation		

0	1	2	3	4	5	6	7	8	9	10	11 x
1	-10	-10	-10	-10	-10	-10	-10	-10	-10	-10	-10
2	-4	-6,80103	-7,7837	-8,13495	-8,24168	-8,25844	-8,24168	-8,13495	-7,7837	-6,80103	-4
3	-4	-5,42042	-6,19883	-6,51444	-6,57331	-6,5504	-6,57331	-6,51444	-6,19883	-5,42042	-4
4	-4	-4,68181	-5,07675	-5,15066	-4,98672	-4,79654	-4,98672	-5,15066	-5,07675	-4,68181	-4
5	-4	-4,23009	-4,2757	-4,02472	-3,42639	-2,66233	-3,42639	-4,02472	-4,2757	-4,23009	-4
6	-4	-3,96284	-3,77124	-3,24615	-2,03177	1	-2,03177	-3,24615	-3,77124	-3,96284	-4
7	-4	-3,85002	-3,60028	-3,15686	-2,45453	-1,65803	-2,45453	-3,15686	-3,60028	-3,85002	-4
8	-4	-3,83696	-3,623	-3,32647	-2,97148	-2,72305	-2,97148	-3,32647	-3,623	-3,83696	-4
9	-4	-3,8748	-3,72831	-3,55453	-3,38188	-3,29119	-3,38188	-3,55453	-3,72831	-3,8748	-4
10	-4	-3,93393	-3,86092	-3,78144	-3,71032	-3,67796	-3,71032	-3,78144	-3,86092	-3,93393	-4
11	-4	-4	-4	-4	-4	-4	-4	-4	-4	-4	-4
y											

Figure 3. Result of the thermo-technical calculation of a multilayered concrete coated pipe in permafrost soil.

The formulated mathematical model allows predicting a change in the temperature field around the buried oil pipeline with sufficient reliability, not only for a homogeneous soil, but also for soil layers with different thermal properties, to consider the change in the density of the soil skeleton as a function of temperature changes with a change in season as well as external factors (foundations of buildings and engineering communications) [10].

3. Conclusion

The implementation of predictive heat engineering calculations of the interaction of the buried oil pipeline with the permafrost soils is an important part of the design process, allowing making justified technical decisions aimed at increasing the reliability and lifetime of the facility.

A comprehensive approach to the selection of the best technical solutions, including the mathematical modeling of the thermal interaction of the pipe-ground system, allows minimizing costs without reducing the efficiency and reliability of the systems.

References

[1] Zemenkov Yu D, Moiseev B V, Ilyukhin K N and Nalobin N V 2014 Mathematic simulation of the interaction of surface pipelines with the environment *News of Higher Educational Institutions. Oil and Gas* 2 pp 51-6

[2] Aksyonov B G, Koryakina S V, Moiseev B V, Shapoval A F and Stepanov O A 2012 Simulation of the vibration of the freeze-thaw boundary in soils and external building structures *Modern Problems of Science and Education*

[3] Gubarkov A A, Idrisov I R and Kirillov A V 2014 Activation of cryogenic processes during the construction of the Bovanenkovo *News of Higher Educational Institutions. Oil and Gas* 2 (Ukhta) pp 6-11

[4] Golubin S I 2011 *Scientific and Methodological Basis for the Forecast of Interaction of Underground Gas Pipelines with Saline Ppermafrost on the Yamal Peninsula* Abstract of PhD thesis of geological and mineralogical sciences (Moscow: Gazprom VNIIGAZ) 24 p

[5] Gorelik J B, Shabarov A B and Sysoyev Yu S The dynamics of frozen ground melting in the influence zone of two wells *Earth's Cryosphere* 12 (1) p 59

[6] Demchenko N P and Krapivskiy E I 2010 Geophysical monitoring of underground pipelines technical state *4th Intern. Conf. and Exhibition: New Discoveries through Integration of Geosciences* (Saint Petersburg)

[7] Grigoriev B V, Chistyakova N F and Shabarov A B 2010 Thermal conductivity of rocks of the

West Siberian oil and gas basin *Bulletin of the Tyumen State University, Physico-Mathematical Simulation. Oil, Gas, Power* 6 pp 19-27

[8] 2009 *Unified Technological Calculations of the Objects of the Main oil Pipelines and Oil Product Pipelines* RD-75.180.00-KTN-198-09 (Moscow) 204 p

[9] 2004 *Thermal Insulation of Equipment and Pipelines* SNIP 41-03-2003 (Moscow: Gosstroy RF) 29 p

[10] 1977 *Handbook of Construction on Frozen Soils* ed Dokuchaev V V (Leningrad: Stroyizdat) 552 p

Assessment of geo-cryological conditions in the design and operation of pipelines in the Arctic zone of the Russian Federation

V V Golik, B V Moiseev, Y D Zemenkov and E N Kabes

Industrial University of Tyumen, 38 Volodarskogo st., Tyumen, 625000, Russia

E-mail: darkpaint@mail.ru

Abstract. The paper presents the methodology of calculations and technical justification for the use of new equipment. The section of the operating oil pipeline in the part of monitoring of heat engineering processes is considered. The main results of thermo-technical calculations of the pipeline section were described and the aureoles of thawing of the permafrost soil were simulated. As an example for replacement, a multilayer pipe with foamed polyurethane insulation and concrete casing was considered because of its better characteristics compared to the design one.

1. Introduction

The construction and operation of pipeline transport facilities for oil and oil products in the Arctic zone of the Russian Federation in which permafrost soils are widespread leads to significant changes in geo-cryological conditions. Progression of negative processes affecting the pipelines stability laid by underground method arises even with the slightest changes in the heat balance of the soil surrounding.

2. Methodology

The design of pipelines in the regions of the Far North and the Arctic zone of the Russian Federation is carried out in accordance with the current regulatory documents. When planning, it is necessary to provide for measures aimed at ensuring long-term and reliable operation of the pipelines to be laid, as well as to develop measures and select materials to minimize the negative impact on the soil and the environment as a whole [1].

To assess the temperature state of soils at the design stage of pipelines in the Arctic zone, thermal engineering calculations are carried out, specified by regulatory documents.

During the work on this subject, the existing normative documents [2] and calculation methods presented in the papers of the authors G.V. Porkhaev, E.Ya. Sokolov, A.F. Shapoval, and N.V. Nalobin were analyzed to increase the efficiency and accuracy of the calculated parameters and display the heat transferring process from the pipeline to the environment by software calculation [3-5].

On the basis of the department "Transport of Hydrocarbon Resources" of the Industrial University of Tyumen, studies have been carried out to simulate the thermal state of pipelines, their behavior in the permafrost soils of the Arctic zone and to monitor the reliability characteristics of pipelines laid by the underground method.

Under the guidance of Professor B.V. Moiseev, a methodology has been developed for calculating and modeling the thermal state of the pipeline in permafrost.

The object of the study was a section of 77 km – 150 km of the Yarudeyskoye oil field located in the Nadym district of the Yamal-Nenets Autonomous District.

This field and the pipeline section were selected according to the following criteria:

- climatic zone with permafrost soil;
- soil composition;
- the presence of sites lying in the swampy terrain;
- pipeline laying by the underground method;
- pipeline defect - buckling of the pipeline (Figure1).

Figure 1. Defective pipeline section.

Table 1. Climatic characteristics of the studied area.

Characteristic	Nadym
Average annual outside air temperature, °C	-6.6
Average annual temperature of the summer period, °C	10
Average annual temperature of the winter period, °C	-14.9
Absolute maximum of the outdoor temperature, °C	35
Absolute minimum of the outside temperature, °C	-60
Average amplitude of temperature change, °C	24.9
Number of days in a year with a negative temperature	229
Average wind speed (January/July) m/s	4.5/4.0
Maximum wind speed m/s	5.0

Deposits in this area are represented by interstratification of clay-loamy rocks, with interlayers and individual lenses of fine, medium-sized and silty sandy loams and sandy loams. Deposits represent a complex mixture of marine, glacial-marine formations. The revealed thickness of clays reaches 15 m, that of loam - 13 m, that of sandy loam - 8 m and that of sand up to 6 m. The soils, drilled in April and May, are mostly in permafrost condition [6-8].

Based on the methodology developed by the team, the following tasks were performed:

- in accordance with the methodology developed, the thermo-technical calculation of the oil pipeline on the section of 77 km-150 km was carried out;
- analysis of calculations and selection of materials for the elimination of a defect were performed;
- software product for calculating the thermal properties of the soil was developed.

With the known climatic data, soil composition and technical characteristics of the pipes, it was necessary to determine heat losses in this pipeline section:

$$q = \pi k \,(t_v - t_n), \quad (1)$$

where k – coefficient of heat transfer from oil to the soil, t_v – temperature of the pumped medium, t_n – ambient temperature of the surrounding soil.

To calculate the heat transfer coefficient, it is necessary to calculate the actual thermal resistance for each layer of the insulation coating of the tested pipeline:

$$R_{in} = \frac{1}{2\pi\lambda_{in}} \ln \frac{D_n}{D_v}, \quad (2)$$

where D_n, D_v – the outer and inner diameters of the insulation layer, respectively, m; λ_{in} – thermal conductivity of the insulating layer at an average insulation temperature.

Based on the calculations, the results are shown in Table 2 and Figures 2, 3. A new pipe was selected for this type of terrain and climate. It was namely multi-layer concrete coated pipe of the company "BT SVAP" ZUB Ballast having similar capacity, but meeting all the requirements for SP 86.13330.2014 Trunk pipelines.

Table 2. Design parameters of the pipelines under investigation.

Dimensions	Units of measurement	Multilayered concrete coated oil pipeline "ZUB Ballast"	Oil pipeline with foamed polyurethane insulation
Outside diameter d	m	0.639	0.503
Heat transfer coefficient k	W/m*K	0.176	0.599
Thermal resistance R	m*K/W	5.676	1.658
External temperature of the pipe t_{sttr}	°C/K	0.95/273.95	17/290
Initial product temperature on the section T_n	°C/K	55/328	
Final temperature on the section T_k	°C/K	51/327	43/316

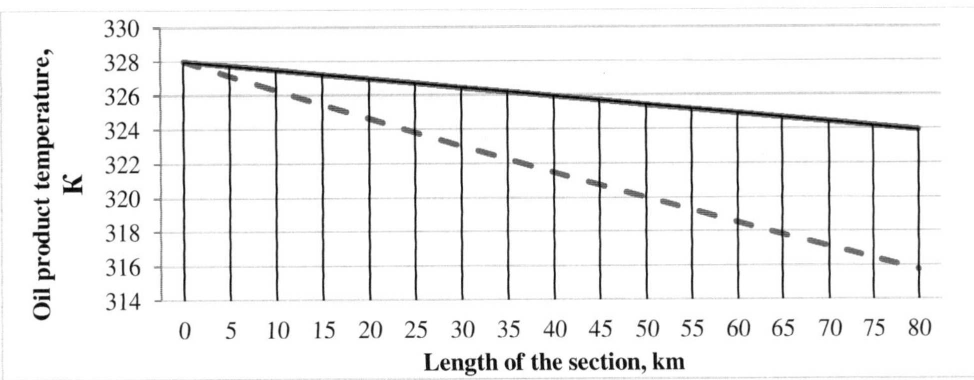

Figure 2. Temperature drop along the length of the oil pipeline:
—— Multilayered concrete covered pipe "ZUB Ballast";
- - - - Oil pipeline with foamed polyurethane insulation.

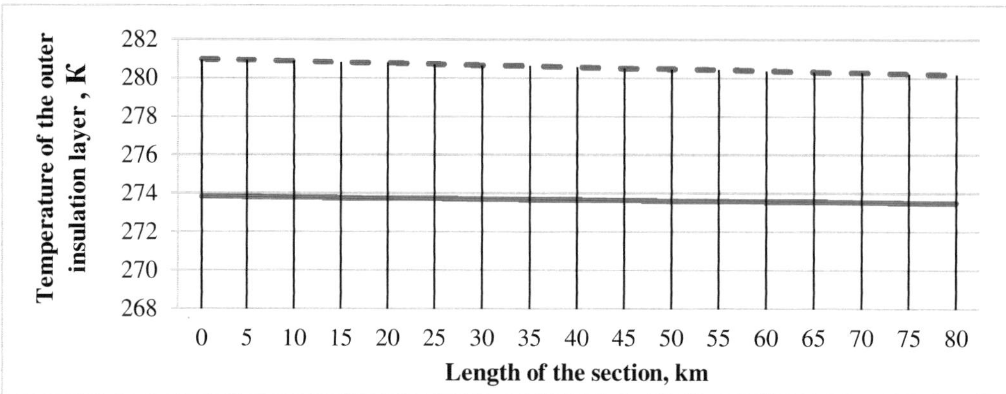

Figure 3. Temperature of the outer layer of insulation:
—— Multilayered concrete covered pipe "ZUB Ballast";
- - - - Oil pipeline with foamed polyurethane insulation.

3. Conclusion

Thus, the performance of heat engineering calculations of the interaction of an oil pipeline laid by underground method with permafrost soils is an important part of the design and operation process which allows making justified engineering decisions aimed at improving the reliability and durability of the facility with minimal environmental impact [9-10].

An integrated approach to the selection of optimal design solutions, including the simulation of thermal interaction in the pipe-ground system, allows the costs optimization without reducing reliability requirements.

The use of the latest software complexes in the simulation and creation of a specialized software product allows the automation of the calculation process and acceleration of the equipment selection.

References

[1] Moiseev B V 1998 *Increase in the Efficiency of the Heat Supply System in the Oil and Gas Producing Areas of Western Siberia* Author's thesis PhD in Tech. Sciences (Tyumen) 66 p

[2] RD-75.180.00-KTN-198-09 *Unified Technological Calculations of the Objects of Main Oil Pipelines and Oil Product Pipelines*

[3] Golubin S I 2011 *Scientific and Methodological Basis for the Forecast of Interaction of Underground Gas Pipelines with Saline Permafrost on the Yamal Peninsula* Author's thesis PhD in Geological and Mineralogical Sciences (Moscow: Gazprom VNIIGAZ) 24 p

[4] Tabunschikov Y 1993 *Mathematical Models of Thermal Conditions in Buildings* (USA: CRC Press)

[5] Novoselov V F and Muftakhov E M 1996 *Technological Calculation of Oil Pipelines* (Ufa: UGNTU) 43 p

[6] Zemenkova M, Shalay V, Zemenkov Y and Kurushina E 2016 Improving the efficiency of administrative decision-making when monitoring reliability and safety of oil and gas equipment *MATEC TPACEE 2016*

[7] Zemenkov Yu D, Shalay V V and Zemenkova M Yu 2015 Expert systems of multivariable predictive control of oil and gas facilities reliability *Proc. Eng.* **113** pp 312-5

[8] Kurushina E V 2014 On regularities of economic dynamics in times of crisis *Canadian J.of Sc., Educ. and Culture* P2 No. 2 (6) pp 378-84

[9] Mamadaliev R A, Kuskov V N, Zemenkov U D and Popova A A 2015 Influence of high-concentrated heat sources on alloying elements transition into the weld metal *Appl. Mech. and Mater.* **770** pp19-22

[10] Gorelik J B, Shabarov A B and Sysoyev Yu S 2016 The dynamics of frozen ground melting in the influence zone of two wells *Earth's Cryosphere* 12 (1) pp 59

Modelling the parameters of emptying an inclined pipeline in the safety management under demanding conditions

M Y Zemenkova[1], A A Gladenko[2] and Y D Zemenkov[1]

[1] Industrial University of Tyumen, 38 Volodarskogo str., Tyumen, 625000, Russia
[2]Omsk State Technical University, 11 Mira Prospekt, Omsk, 644050, Russia

E-mail: yd_zemenkov@mail.ru, gladenko1961@yandex.ru

Abstract. The work describes the method of estimating the volume of oil products during outflow and emergency discharge. Before removing the defective area, the pipeline is freed from oil products to ensure safety of repair and recovery works. The method has been tested on the main pipelines of the Tyumen Oblast. Based on the results of calculations and experiments, it was established that for the implementation of expert estimates, the error in the calculations by the methods proposed is satisfactory.

1. Introduction

The damage to the environment ultimately depends on the amount of potential oil flow and the location of the damage. The availability of data on the magnitude of the potential flow at various sections of the pipeline and the likelihood of failure on them makes it possible, at the design stage, to provide for the most effective measures to reduce damage to nature in the event of oil product leakage. At present, oil sector enterprises pay great attention to oil pipeline integrity control; decision support systems such as leak detection systems, integrity monitoring of long-distance facilities, etc. are being created.

However, the issues of instrument monitoring, detection, assessment and prediction of leakage, location, development of accurate operational calculations remain relevant [1-10].

Gravity drainage of the shutoff pipeline section with hydraulic slopes greater than 0.001 is typical for most main pipelines and occurs at full cross section. After the gravity drainage is stopped, the product remaining in the pipe is pumped out with pumps. The longest operations (percentage of the total time for the accident elimination) are: emptying of the pipeline - up to 35%; developing of a pit to collect oil - up to 9%; sealing of the internal cavity of the pipeline and preparation for conducting hot works - up to 16%; preparation of a new coil and its installation on a pipeline - up to 15% [1]. As practice shows, the duration of emptying the damaged section shut off by line valves varies widely and can amount to more than 50% of the total duration of the restoration work.

2. Materials and methods

At different times, research for the prediction and assessment of leakage and outflow models was conducted under the guidance of P.I. Tugunov, R.A. Aliev, M.V. Lurie, A.G. Gumerov, S.V. Kuklev, G.D Rosenberg, V.N. Antipyev, V.A. Polyakov, A.S. Shumailov, T.M. Bashty, S.S. Rudnev, B.B. Nekrasov, V.A. Fisher, and other well-known scientists.

At present, there are methods for calculating the outflow from pipelines, each of which has a specific physical and mathematical formulation, advantages and disadvantages. For example, the authors Donets

K.G. and Chernikin V.I. considered the process of gravity drainage of pipelines in case of accidents involving rupture of pipelines and termination of pumping [2]. They described the process of outflow by the Bernoulli equation in integral form, considering the following features:

- at the saddle point, the continuity of the liquid flow breaks, so a pressure equal to the saturated vapor pressure is established above the oil level p_s;
- there is friction of the liquid against the wall of the pipe;
- the inertial term is considered.

Initial conditions for differential equations are considered in the publication of Voloshina A.P. in the following form:

$$\text{at } t=0{:}\; Z=H_1;\; Z^{\cdot}=0, \tag{1}$$

and the continuity equation is written considering the profile of the route:

$$Qdt = \left(\frac{\pi D^2}{4sin\alpha}\right) \cdot dZ, \tag{2}$$

where D, α - diameter, angle of inclination of the pipeline to the horizon; Q - volumetric flow of liquid in the pipeline; Z – fluid head.

Under these conditions, it is found that the velocity of the fluid in the pipeline obeys the condition

$$\vartheta = \frac{g \cdot D^2 \cdot \sin\alpha}{16 \cdot \nu} \cdot \left[1 - \exp\left(-\frac{32 \cdot \nu \cdot t}{D^2}\right)\right]. \tag{3}$$

A significant disadvantage of this technique is the fact that the velocity of the liquid in the pipeline is not related to the flow rate of the liquid from the hole, therefore, these dependencies do not include the flow coefficient that considers the actual flow rate. Here, it can be emphasized that when the product flows through an opening in an inclined pipeline, an unsteady motion is observed with a constant decrease in velocity and a change in the flow pattern. In addition, the authors considered only the laminar flow pattern of the liquid in the pipeline, which also limits the scope of application of the results obtained in practical calculations. The results of calculations [2-4 and others] showed that with an increase in the diameter of the pipeline, with an equal ratio of the area of the discharge hole to the cross-sectional area of the pipe and other equal conditions, the duration of emptying the inclined section decreases. This is because the friction pressure loss along the length of the pipeline, per unit volume of the outflowing product, decreases with increasing diameter, and the velocity, respectively, increases.

In [2], the problem of emptying a pipeline in case of sudden damage was considered. The main question of the study is the question of determining the time of the outflow of fluid from the damaged section of the pipe. The calculation was made for a pipeline 10 km long, 500 mm in diameter with a fixed 1:10 ratio of the hole area to the pipe area. It was assumed that the pipeline was filled with gasoline with a density of ρ = 760 kg/m^3. As a result of the calculations given by the author, it turned out that in 25 minutes about 210 m^3 of liquid flows out of the pipeline through the damage.

The works of Biryukov A.E., Turkin V.N., Bobrovskiy S.A., Gross S.A., Yanov B.G. deal with determining the volume of the emptying section or the time of emptying. But they consider the process simplistically, without considering a number of factors, for example, the wave structure of the fluid motion through the rupture, the time variability of the fluid flow coefficient, the one-dimensionality of the fluid motion in the pipeline, etc. In [7] various possible pipeline conditions are given, for which semi-analytical methods of leakage calculation are recommended. In practice, in case of accidents on main pipelines, the amount of oil removed from the pipeline at the third stage is determined by actual data, based on the method of emptying and available technical means.

Before removing the defective area, the pipeline is freed from oil products in order to ensure safety of repair and recovery works. The volume of the leaked and pumped out product can be determined by summing the corresponding volumes of descending sections of the accident site. Part of the descending sections is freed from the product completely. These sections are located along the profile of the route above the rupture site and their volume is equal to

$$V = 0.25 \cdot \pi \cdot \sum_{i=1}^{n} D_i^2 \cdot l_{in}, \tag{4}$$

where n - number of descending sections.

The oil product is discharged from part of the sections partially depending on the angle of their inclination to the horizon and the location relative to the point of discontinuity. For such sites, the volume of leaked oil is determined as the volume of an irregular truncated cylinder [8].

We derive a formula for calculating such volumes by introducing the following notation, using a three-dimensional coordinate system (Figure 1).

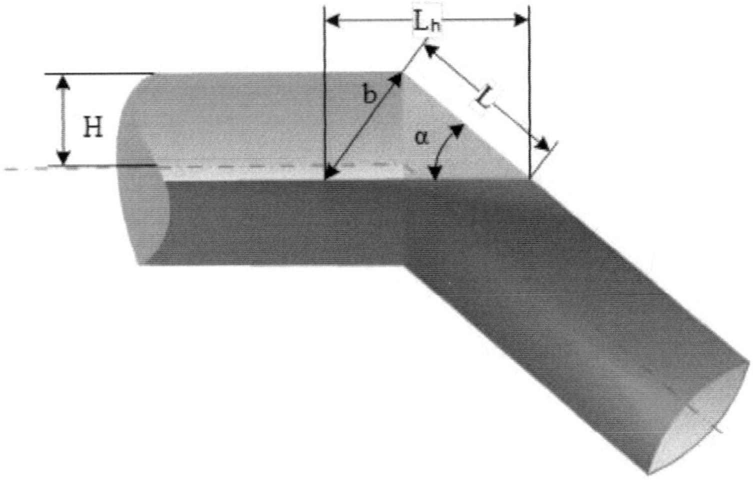

Figure 1. Design scheme of the pipeline with the angle of inclination α with respect to the horizon α - angle of inclination of pipeline with respect to the horizon, deg; h - height of the section, m; Lh - length of the horizon, m; L - length of the pipeline section, m.

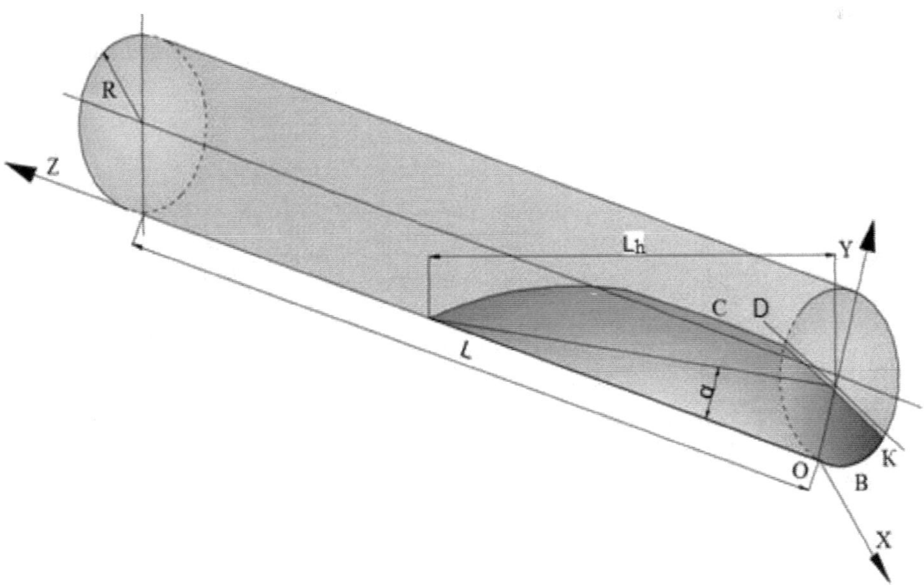

Figure 2. Design scheme of the pipeline: three-dimensional diagram of a damaged pipeline; α - angle of inclination of pipeline with respect to the horizon, deg; h - height of the section, m; Lh - length of the horizon, m; L - length of the pipeline section, m.

From geometric constructions it follows that:

$$b = L_h \cdot \sin\alpha \tag{5}$$

$$L = L_h \cdot \cos\alpha \tag{6}$$

$$b = L_{pl} \cdot tg\,\alpha \quad . \tag{7}$$

To determine the coordinates of the points B and C through which the secant plane passes, we write the following equations:

$x^2 + (y - R)^2 = R^2$ - circumferential equation, (8)

$Z = 0$ - equation of the XOY plane, (9)

$y = b$ - equation of a plane intersecting XOY along a straight line KD. (10)

We conduct the transformations (8):

$$(y - R)^2 + x^2 - R^2 = 0, \tag{11}$$

$$\text{or} \quad y^2 - 2Ry + x^2 = 0. \tag{12}$$

We solve the last equation:
for $R \leq b \leq 2R$ we obtain the root:

$$y = R + \sqrt{R^2 - x^2}, \tag{13}$$

and for $0 \leq b < R$:

$$y = R - \sqrt{R^2 - x^2}. \tag{14}$$

Let us consider the root (13). It follows from (11) that

$$R + \sqrt{R^2 - x^2} = b,$$

$$\text{or} \quad \sqrt{R^2 - x^2} = b - R. \tag{15}$$

Let us square the left and right parts:

$$R^2 - x^2 = (b - R)^2.$$

After the transformations, we obtain

$$x_{1,2} = \pm\sqrt{R^2 - (b - R)^2}. \tag{16}$$

Now we can write the coordinates of 3 points through which the horizontal secant plane passes:

$$\begin{aligned} &m.A \quad (0,0,L), \\ &m.B \quad \left(\sqrt{R^2 - (b-R)^2}, b, 0 \right), \\ &m.C \quad \left(-\sqrt{R^2 - (b-R)^2}, b, 0 \right). \end{aligned} \tag{17}$$

We solve the system of equations:

$$\begin{vmatrix} y_A & z_A & 1 \\ y_e & z_B & 1 \\ y_C & z_C & 1 \end{vmatrix} \cdot x + \begin{vmatrix} z_A & x_A & 1 \\ z_e & x_B & 1 \\ z_C & x_C & 1 \end{vmatrix} \cdot y + \begin{vmatrix} x_A & y_A & 1 \\ x_e & y_B & 1 \\ x_C & y_C & 1 \end{vmatrix} \cdot z = \begin{vmatrix} x_A & y_A & z_A \\ x_e & y_B & z_B \\ x_C & y_C & z_C \end{vmatrix}. \tag{18}$$

The determinant for x

$$\Delta x = \begin{vmatrix} 0 & L & 1 \\ b & 0 & 1 \\ b & 0 & 1 \end{vmatrix} = L \cdot b - L \cdot b = 0. \tag{19}$$

The determinant for y

$$\Delta \mathbf{y} = \begin{vmatrix} L & 0 & 1 \\ 0 & \sqrt{R^2-(b-R)^2} & 1 \\ 0 & -\sqrt{R^2-(b-R)^2} & 1 \end{vmatrix} = L\cdot\sqrt{R^2-(b-R)^2} + \tag{20}$$

$$+L\cdot\sqrt{R^2-(b-R)^2} = 2L\cdot\sqrt{R^2-(b-R)^2}.$$

The determinant for z

$$\Delta z = \begin{vmatrix} 0 & 0 & 1 \\ \sqrt{R^2-(b-R)^2} & b & 1 \\ -\sqrt{R^2-(b-R)^2} & b & 1 \end{vmatrix} = b\cdot\sqrt{R^2-(b-R)^2} +$$

$$+b\cdot\sqrt{R^2-(b-R)^2} = 2b\cdot\sqrt{R^2-(b-R)^2}. \tag{21}$$

The main determinant of the system (18)

$$\Delta = \begin{vmatrix} 0 & 0 & L \\ \sqrt{R^2-(b-R)^2} & b & 0 \\ -\sqrt{R^2-(b-R)^2} & b & 0 \end{vmatrix} = b\cdot L\cdot\sqrt{R^2-(b-R)^2} +$$

$$+b\cdot L\sqrt{R^2-(b-R)^2} = 2b\cdot L\sqrt{R^2-(b-R)^2}. \tag{22}$$

We write the equation of the plane:

$$2L\cdot\sqrt{R^2-(b-R)^2}\cdot y + 2b\cdot\sqrt{R^2-(b-R)^2}\cdot z = 2Lb\cdot\sqrt{R^2-(b-R)^2} \tag{23}$$

Dividing the left and right sides by $2\cdot\sqrt{R^2-(b-R)^2}$, we obtain

$$L\cdot y + b\cdot z - L\cdot b = 0, \tag{24}$$

or

$$z = \frac{L\cdot(b-y)}{b}. \tag{25}$$

We proceed to determine the volume of a truncated cylinder:

$$\mathbf{V} = \iint_{ABCD} z\cdot dx\cdot dy = \int_0^b dy \int_{-\sqrt{R^2-(y-R)^2}}^{\sqrt{R^2-(y-R)^2}} \frac{(L\cdot b - L\cdot y)}{b}\cdot dx =$$

$$= \int_0^b \frac{(L\cdot b - L\cdot y)}{b}\cdot 2\cdot\sqrt{R^2-(y-R)^2}\cdot dy =$$

$$= 2\cdot L\cdot\int_0^b \cdot\sqrt{R^2-(y-R)^2}\cdot dy - \frac{2L}{b}\cdot\int_0^b \cdot\sqrt{R^2-(y-R)^2}\cdot dy. \tag{26}$$

We transform the second integral:

$$\int_0^b y\cdot\sqrt{R^2-(y-R)^2}\cdot dy = \int_0^b \cdot(y-R+R)\cdot\sqrt{R^2-(y-R)^2}\cdot dy =$$

$$= R\cdot\int_0^b \cdot\sqrt{R^2-(y-R)^2}\cdot dy + \int_0^b \cdot(y-R)\cdot\sqrt{R^2-(y-R)^2}\cdot dy. \tag{27}$$

We obtain

$$\mathbf{V}=2L\cdot\int_0^b\cdot\sqrt{R^2-(y-R)^2}\cdot dy-\frac{2L}{b}\cdot R\int_0^b\cdot\sqrt{R^2-(y-R)^2}\cdot dy-$$
$$-\frac{2L}{b}\cdot\int_0^b\cdot\sqrt{R^2-(y-R)^2}\,(y-R)\cdot dy. \quad (28)$$

We combine in this expression the first two terms:

$$\mathbf{V}=\frac{2\cdot Lb-2\cdot LR}{b}\cdot\int_0^b\sqrt{R^2-(y-R)^2}\cdot dy-\frac{2L}{b}\cdot\int_0^b\sqrt{R^2-(y-R)^2}\cdot(y-R)\cdot dy. \quad (29)$$

Using table integrals, we obtain:

$$\mathbf{V}=-\frac{(2L(b-R))}{b}\cdot\left\{\frac{y-R}{2}\cdot\sqrt{R^2-(y-R)^2}+\frac{R^2}{2}\cdot\arcsin\frac{(y-R)}{R}\right\}\Big|_0^b-$$
$$-\frac{2L}{b}\cdot\left\{-\frac{1}{3}\left[R^2-(y-R)^2\right]^{3/2}\right\}\Big|_0^b \quad (30)$$

We substitute the limits:

$$\mathbf{V}=\frac{2L\cdot(b-R)}{b}\left\{\frac{(b-R)}{2}\cdot\sqrt{R^2-(b-R)^2}+\frac{R^2}{2}\arcsin\left(\frac{(b-R)}{R}\right)+\frac{R^2}{2}\cdot\frac{\pi}{2}\right\}+$$
$$+\frac{2L}{b}\left\{\frac{1}{3}\cdot\left[R^2-(b-R)^2\right]^{3/2}\right\}. \quad (31)$$

Thus, using table integrals when solving (28), it is possible to obtain the volume of product to be pumped out:

$$\mathbf{V}=\frac{2L}{b}\left\{\frac{(b-R)}{2}\left[(b-R)\sqrt{R^2-(b-R)^2}+R^2\arcsin\left(\frac{b-R}{R}\right)+\frac{\pi\cdot R^2}{2}\right]+\frac{1}{3}\left[R^2-(b-R)^2\right]^{3/2}\right\}. \quad (32)$$

We obtain a solution for another root (2.50) ($0\le b<R$). Using equations (14), we obtain

$$R+\sqrt{R^2-x^2}=b, \quad (33)$$

or $\sqrt{R^2-x^2}=R-b$.

Let us square the left and right parts:

$$R^2-x^2=(R-b)^2 \quad (34)$$

After the transformations, we obtain

$$x_{1,2}=\pm\sqrt{R^2-(R-b)^2}. \quad (35)$$

Arguing in the same way as for the previous root (13), we obtain the equation of the plane (25).

We now turn to determining the volume of a truncated cylinder for the root (14):

$$\mathbf{V}=\iint_{ABCD}z\cdot dx\cdot dy=\int_0^b dy\int_{-\sqrt{R^2-(R-y)^{22}}}^{\sqrt{R^2-(R-y)^2}}\frac{(L\cdot b-L\cdot y)}{b}\cdot dx=\int_0^b\frac{(L\cdot b-L\cdot y)}{b}\cdot 2\cdot\sqrt{R^2-(R-y)^2}\cdot dy=$$
$$=2\cdot L\cdot\int_0^b\cdot\sqrt{R^2-(R-y)^2}\cdot dy-\frac{2L}{b}\cdot\int_0^b\cdot\sqrt{R^2-(R-y)^2}\cdot dy. \quad (36)$$

We obtain

$$\mathbf{V} = 2L\cdot\int_0^b \cdot\sqrt{R^2-(R-y)^2}\cdot dy - \frac{2L}{b}\cdot R\int_0^b \cdot\sqrt{R^2-(R-y)^2}\cdot dy -$$

$$-\frac{2L}{b}\cdot\int_0^b \cdot\sqrt{R^2-(R-y)^2}\,(y-R)\cdot dy. \quad (37)$$

We combine in this expression the first two terms:

$$\mathbf{V} = -\frac{2\cdot L\cdot b - 2\cdot LR}{b}\cdot\int_0^b \sqrt{R^2-(R-y)^2}\cdot d(R-y) -$$

$$-\frac{2L}{b}\cdot\int_0^b \sqrt{R^2-(R-y)^2}\cdot(R-y)\cdot d(R-y). \quad (38)$$

We use table integrals:

$$\mathbf{V} = -\frac{(2L(b-R))}{b}\cdot\left\{\frac{R-y}{2}\cdot\sqrt{R^2-(R-y)^2} + \frac{R^2}{2}\cdot\arcsin\frac{(R-y)}{R}\right\}\Bigg|_0^b -$$

$$-\frac{2L}{b}\cdot\left\{-\frac{1}{3}\left[R^2-(R-y)^2\right]^{3/2}\right\}\Bigg|_0^b \quad (39)$$

We substitute the limits:

$$\mathbf{V} = \frac{2L\cdot(b-R)}{b}\cdot\left\{\frac{(b-R)}{2}\cdot\sqrt{R^2-(R-b)^2} - \frac{R^2}{2}\cdot\arcsin\left(\frac{R-b}{R}\right) + \frac{R^2}{2}\cdot\frac{\pi}{2}\right\} +$$

$$+\frac{2L}{b}\left\{\frac{1}{3}\cdot\left[R^2-(R-b)^2\right]^{3/2}\right\}. \quad (40)$$

Thus, using table integrals when solving (36), it is possible to obtain the volume of product to be pumped out:

$$\mathbf{V} = \frac{2L}{b}\left\{\frac{(b-R)}{2}\left[(b-R)\sqrt{R^2-(R-b)^2} - R^2\arcsin\left(\frac{R-b}{R}\right) + \frac{\pi\cdot R^2}{2}\right] + \frac{1}{3}\left[R^2-(R-b)^2\right]^{3/2}\right\}. \quad (41)$$

Checking the formulas (for $b=R$), we obtain a well-known formula for the volume:

$$\mathbf{V} = \frac{2\cdot L}{b}\cdot\frac{1}{3}\cdot\left(R^2\right)^{3/2} = \frac{2}{3}\cdot L\cdot R^2. \quad (42)$$

The obtained formulas (32) and (41) cover the entire range of values of the quantity b: $(0 < b < 2R)$.

Formulas (32) and (41) are fairly simple and convenient for practical use in express analyzes in the field, and are also easily implemented using software packages.

3 Conclusion

The method has been tested on the main pipelines of the Tyumen region. Theoretical assumptions about an increase in the error of calculations with an increase in the total length of the controlled sections L, with an increase in emergency x_a, after emergency distances $(L - x_a)$, etc. have been experimentally confirmed. Based on the results of calculations and experiments, it was established that for the implementation of expert estimates, the error in the calculations by the methods proposed is satisfactory.

References

[1] Sviridova L T et al 1989 Emergency repair of main oil pipelines in swamps *Transport and Storage of Oil and Petroleum Products* (Moscow: VNIIOENG) 60 p

[2] Donets K G and Chernikin V N 1963 Gravitational emptying of the pipeline from viscous oils and oil products *Transport and Storage of Oil* 11 pp 3-6

[3] Shumailov A S et al 1981 Controlling leaks of oil and petroleum products on main pipelines during operation *Transport and Storage of Oil and Petroleum Products* 10 (Moscow: VNIIOENG) 80 p

[4] Shumailov A S, Gumerov A G and Moldavanov O I 1992 *Diagnostics of Main Pipelines* (Moscow: Nedra) 291 p

[5] Biryukov A E and Turkin V N 1983 Determination of the time of emptying the pipeline during its rupture on reclamation systems *Hydrological and Melioration Calculations and Characteristics of Some Areas of Siberia and Kazakhstan Coll. of Sc. Papers* (Omsk) pp 58-63

[6] Bobrovsky S.A. Determination of the time of idling oil pipelines in the elimination of accidents / Gubkin MINKhiGP Moscow: 1963, issue.45-p.181.

[7] Ministry of Fuel and Energy 1996 *Methodology for Determining Damage to the Natural Environment in Case of Accidents on Main Oil Pipelines* (Moscow: TransPress) 66 p

[8] 2008 *Monitoring of Hydrodynamic and Technical Characteristics of Pipeline Systems* ed Zemenkov Yu D (Tyumen: Vector Book) 445p

[9] 2018 Safety guide *Methodological Recommendations on the Classification of Man-made Events in the Field of Industrial Safety at Hazardous Production Facilities of the Oil and Gas Industry* Approved by the Order of the Federal Service for Environmental, Technological and Nuclear Supervision of January 24, 2018 No. 29 [Electronic resource]. - Access mode: http://sudact.ru/law/prikaz-rostekhnadzora-ot-24012018-n-29-ob/rukovodstvo-po-bezopasnosti-metodicheskie-rekomendatsii/ (access date: 1.04.2018)

[10] 2016 On choosing multicomponent multiphase system separation progress optimization criteria *IOP Conf. Ser.: Mat. Sc. and Eng.* 154(1)

[11] Chekardovskiy M N, Chekardovskiy S M and Ilyukhin K N 2016 Methods for determining the thermodynamic parameters of gas compressor units of main gas pipelines *IOP Conf. Ser.: Mat. Sc. and Eng.* 154(1)

[12] Gorelik J B, Shabarov A B and Sysoyev Yu S 2016 The dynamics of frozen ground melting in the influence zone of two wells *Earth's Cryosphere* 12 (1) pp 59

Forecasting and managing the reliability of technological equipment of oil and gas pipelines with the use of real-time reserving

M Y Zemenkova[1]

[1]Industrial University of Tyumen, 38 Volodarskogo str., Tyumen, 625000, Russia

E-mail: muzemenkova@mail.ru

Abstract. The paper deals with evaluation of reliability indicators of technological equipment of oil and gas pipelines using graph models. A special category is represented by evaluation techniques that describe complex facilities with a time reserve with the possibility of prompt restoration. Researchers at the Department of Hydrocarbon Transportation are carrying out research to create decision support systems with predictive, warning and reliability assessment functions. The proposed method can be used for systems with a structural and time reserve. When the system operates in real time and determines the exact type and parameters of failure distribution (physical or parametric), the results can be effectively used within the expert system of reliability management.

1. Introduction

Most units of the technological equipment of the oil and gas complex are recoverable complex objects or, formalizing, systems. Mathematical methods of applying graph methods described by the authors in [1, 5-13] provide for separate work with each technical facility. In the case where failure is considered and diagnosed as parametric, the nature of the parameter change should be considered.

The new requirements in the field of industrial safety were approved on January 24, 2018. The order of the Federal Service for Ecological, Technological and Nuclear Supervision No. 29 in the oil and gas industry started a new stage - the safety manual "Methodological recommendations on the classification of man-made events in the field of industrial safety at hazardous production facilities of the oil and gas complex" was approved [4].

Technogenic events in the field of industrial safety are now recommended to be classified on the basis of technological features of the facility, signs of the realization of the hazard of accidents, the severity of consequences into four levels of danger: level 1 - accident; level 2 - incident; level 3 - premise to the incident (hereinafter - premise); level 4 - violations in the system of industrial safety management or deviations of technological parameters, but without exceeding the maximum permissible values, including those registered by remote control. A modern classification requires novel approaches and technology of control, differentiation and management of events. For example, it is now recommended to evaluate the parameters for an anthropogenic event using models corresponding to the specifics of the facility. Let us consider the features of constructing a model on the basis of ensuring the status of working capacity, as the basic identifier for estimation on production.

Modern systems of automatic control of technological processes in the presence of a structural reserve make it possible to switch almost instantly, for example, a main unit into a repair mode with the connection of a backup unit in a "hot" reserve. It is assumed that the time of repair or running-in of the failed unit is carried out for some time, which we will consider as the recovery time. Analysis of regulatory industry documents shows that the usually acceptable time of loss of inoperability is from 8 to 72 hours, in particularly critical cases - about 1 hour. The presence of a replenishable reserve in the simulated system leads to the fact that the failure of an object does not mean simultaneous failure of the system, if the recovery of its operability occurs within the time not exceeding the permissible. Thus, the use of the technology of separate redundancy allows to avoid additional expenditure of material resources for increasing the reliability of elements and the entire facility.

2. Materials and methods

Most types of process equipment have different types of failure distribution, in addition, there is maintenance as a recovery procedure. It is proved that the statistical distribution of failures of complex equipment and systems is not limited to an exponential type of distribution, but requires a refinement in real time [1,2,3]. In this case, methods for assessing the risks of events require a fairly accurate prediction of events.

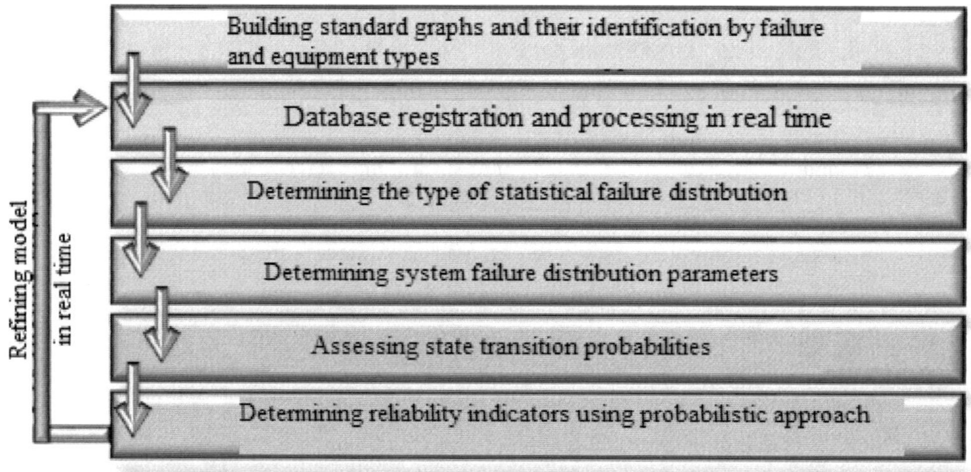

Figure 1. General algorithm for realizing the probabilistic approach when predicting reliability in real time on a structural node of a station or pipeline.

Let us consider the basic model most applicable in assessing the reliability of technological equipment of oil and gas pipelines with a recovery time reserve. The main assumptions used in this reliability assessment algorithm will be the assumption of the exponential nature of the distribution and the total of two basic states of the object - operable (0) and inoperable (1). The basics of constructing mathematical expressions are detailed in [4-13].

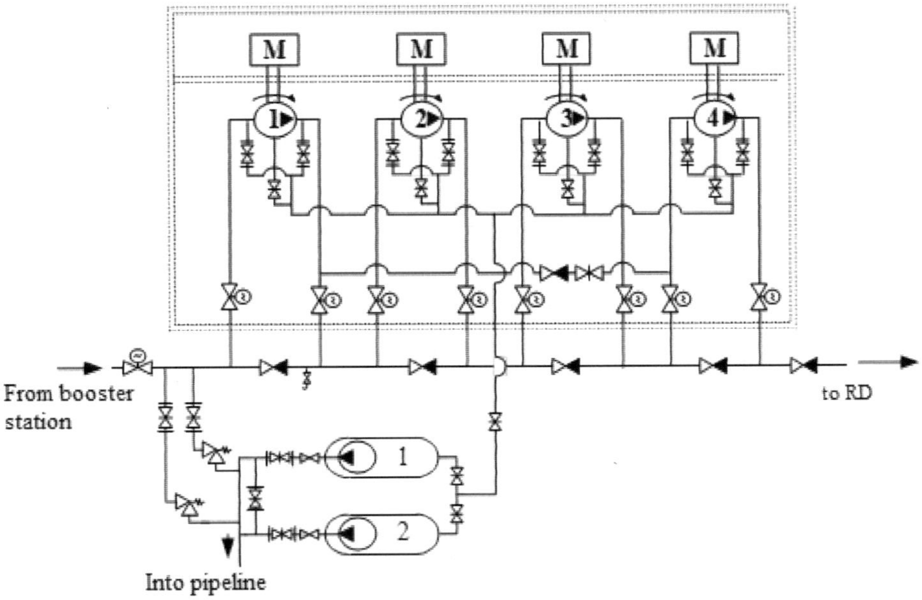

Figure 2. An example of fitting of pump and power units in a pumping station with the possibility of structural redundancy and provision of a time reserve.

According to this feature, two classes of systems are distinguished: first class (the use of only time redundancy) and second class (the sharing of time and structural redundancy). Both models are applicable in the system of operation of pipeline transport facilities for hydrocarbons.

By the time of each next object failure, the time reserve is replenished to the original value regardless of the number of failures, the time for elimination and the time between failures. The distribution law D (t) is assigned, which means that the form of the distribution law D (t) and its parameters remain unchanged with each use of the time reserve. The total recovery time of the object is not limited. Consequently, the restriction on the duration of each recovery of the operability of an object is the only restriction here on the way of using the time reserve.

Different load modes of backup elements, cases of "quick" recovery of the operability of elements, the possibility of switch failures and other factors characterizing the actual functioning of systems.

Let us consider a system that includes an object represented by one structural element, and a recoverable time reserve. The non-failure operating time between adjacent failures tn has the same distribution F(t) as the operating time t_0 before the first failure. The recovery time t_r is $F_r(t)$, independent of the failure location and the previous non-failure operating time of the object. During the repair process, the object completely restores its original properties. The time reserve t_d used in the system is $D(t) = P(t_d < t)$. We assume that F(t), $F_r(t)$ and D(t) have continuous distributions densities.

The functioning of the system is as follows: the object, having worked for a random time t_{n1}, fails and then is recovered at a random time t_{r1}; after recovery of operability it again works for a random time t_{r2}, then it is recovered again for a time t_{r2}, then it is recovered for a time t_{n2} and so on. In view of the assumptions, the values of t_{ni} and t_{ri} are independent, and $P(t_{ni}<t)=F(t)$ and $P(t_{ri}<t)= F_r(t)$, i and j=1,2,..

At the time of each object failure, the time reserve is "switched on". If the recovery of the operability of the object ends before the expiration of the time reserve, it is considered that this object failure violates the normal functioning of the system. The peculiarity of the system under consideration is that short intervals of recovery time for the operability of the object $t_r < t_d$ refer to the useful time.

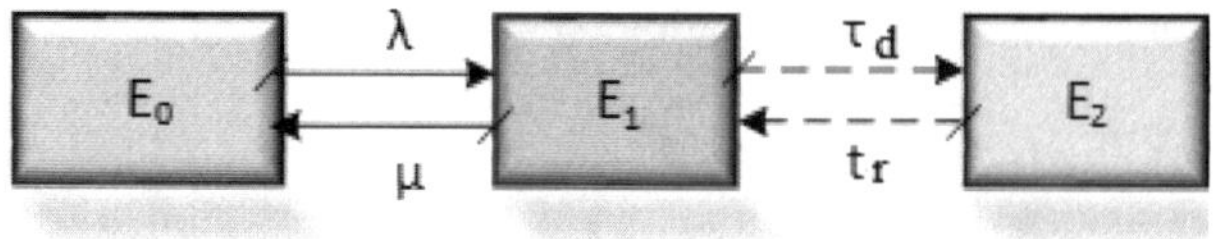

Figure 3. Graphical representation of the process of functioning of an object (a) and system (b) in time.

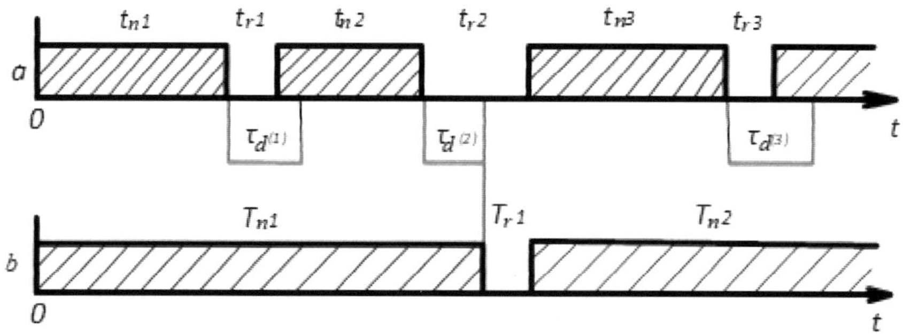

Figure 4. Graphical representation of the process of real-time functioning of an object (a) and system (b) for a node.

Determination of the probability of failure-free operation of the system. The state of the system works in two states: e_0 (the object is operable) and e_1 (the object is inoperable, in recovery for a period of time), state e_2 is inoperable.

We denote by Q_0 (s, t_d) and Q_1 (s, t_d) the Laplace-Stieltjes transformations of time to failure of the system from the states e_0 and e_1, respectively. We can write the following system of equations:

$$\begin{cases} Q_0(s,t_d) = P_{01}(s)Q_1(s,t_d), \\ Q_1(s,t_d) = P_{10}(s)Q_1(s,t_d) + P_{12}(s), \end{cases} \tag{1}$$

Solving the system of equations for Q_0 (s, t_d), we get (2):

$$Q_0(s,t_d) = \frac{P_{01}(s)P_{12}(s)}{1-P_{01}(s)P_{10}(s)} \tag{2}$$

Let us determine the values of the quantities in (2).

From the state e_0, only a transition to the state e_1 is possible, hence $P_{01}(t) = F(t)$, under the condition of exponential distribution, we obtain:

$$P_{01}(s) = \int_0^{\infty} e^{-st} dF(t), \tag{3}$$

And the probability of transition to the state e_1 in time to t is determined as

$$P_{10}(s) = \int_0^{t} [1 - D(t)] dF_B(t). \tag{4}$$

The probability of transition to an inoperable state e2 is calculated as

$$P_{12}(s) = \int_0^{\infty} e^{-st} [1 - F_B(t)] dD(t). \tag{5}$$

Since in the special assumption the distribution is exponential

$$F(t) = 1 - e^{-\lambda t}, \tag{6}$$

$$F_B(t) = 1 - e^{-\mu t}, \tag{7}$$

$$D(t) = 1 - e^{-\gamma t}. \tag{8}$$

The Laplace transform will have the form:

$$LP(t,t_d) = \frac{s+\lambda+\mu+\gamma}{s^2+s(\lambda+\mu+\gamma)+\lambda\gamma}, \tag{9}$$

IOP Conf. Series: Materials Science and Engineering **445** (2018) 012007 doi:10.1088/1757-899X/445/1/012007

$$P(t,t_d) = \frac{1}{C}[Be^{-At} - Ae^{-Bt}], \tag{10}$$
$$A = 0.5(\lambda + \mu + \gamma - C), \tag{11}$$
$$B = 0.5(\lambda + \mu + \gamma + C), \tag{12}$$
$$C = \sqrt{(\lambda + \mu + \gamma)^2 - 4\lambda\gamma} \tag{13}$$

If the time reserve is a non-random, constant value t_d for the probability $P(t, t_d)$ an exact calculation relation can also be found:

$$P(t,t_d) = 1 - \sum_{i=0}^{[t/t_d]-1} \frac{k_{av}^i(1-k_{av})^{i+1}}{(i!)^2} e^{-(1+i)\mu t_d} \cdot \left\{-k_{av}\frac{(2i+1)!}{i+1} + \sum_{j=0}^{i}\left(\frac{i}{j}\right)(i+j)! \times \right.$$
$$\left.[a--(i+1)b]^{i-j} \cdot \left[(-1)^{i+j}\left(1+k_{av}\frac{a-(i+j)b}{i+1-j}\right) - (1-k_{av}\frac{i+j+1}{i+1}e^{-a+b(i+1)}\right]\right\} \tag{14}$$
$$k_g = \mu/(\lambda + \mu) \tag{15}$$
$$a = (\lambda + \mu)t \tag{16}$$
$$b = (\lambda + \mu)t_d \tag{17}$$

Calculation relationships for the probability of failure-free operation $P(t, t_d)$:

$$P(t,t_d) = \exp(-\lambda q t), t \gg t_d \tag{18}$$
$$P(t,t_d) = \begin{cases} 1, t \le t_d \\ exp[-\lambda(t-t_d)(1-F_r(t_d))], t > t_d \end{cases} \tag{19}$$

The probability that the failure of the object will lead to the failure of the entire system is determined as

$$q = P\{t_r > \tau_d\} = \int_0^\infty (1 - F_B(t))dt(t) \tag{20}$$

Time between failures:

$$\overline{T_n}(\tau_d) = \frac{1}{q}(t_n + Mmin(t_r, \tau_d)), \tag{21}$$

The average recovery time of the system is determined as:

$$\overline{T_r}(t_d) = \int_{t_d}^{\infty} \frac{(t-t_d)dF_r(t)}{1-F_r(t_d)} = \frac{\overline{t_r} - \mathrm{Mmin}(t_r, t_d)}{1-F_r(t_d)} \tag{22}$$
$$\overline{T_r}(t_d) = \frac{1}{q}[\overline{t_r} - \mathrm{M}\min(t_r, \tau_d)] \tag{23}$$

The system availability factor is determined by the formula:

$$K_{av}(t_d) = k_{av} + (1 - k_{av})(1 - q) \tag{24}$$

Dependencies (22-24) are designed to determine time between failures, average recovery time, system availability and operation availability factors. The advantage of this algorithm is flexibility and the ability to be complicated in real time. More complex models using different types of distribution are presented by the author in [1].

3. Conclusion

Thus, using the presented calculation technique, it is possible to predict key reliability parameters by the dynamics of the failure rate and the type of distribution. The algorithm and methodology are designed for functioning within the framework of an expert control system for reliability of energy-mechanical equipment of oil and gas pipelines. When the system operates in real time and determines the exact type and parameters of failure distribution (physical or parametric), the results can have different accuracy, complexity, and a list of diagnosed indicators. Based on the monitoring results of the technological facility, not only management solutions can be adopted, but also solutions on constructive system improvement.

References

[1] Zemenkova M Yu 2017 *System Analysis and Technological Monitoring of Reliability and Safety in the Transport and Storage of Hydrocarbons* (Tyumen: IUT) 270 p

[2] Gorelik J B, Shabarov A B and Sysoyev Yu S 2016 The dynamics of frozen ground melting in the influence zone of two wells *Earth's Cryosphere* 12 (1) pp 59

[3] Vlasenko V S, Slesarenko V V, Gulkov A N and Lapshin V D 2015 Preparation of arctic oil and gas condensate deposit formation fluid for storage and transport in the form of hydrocarbon hydrate-containing dispersed system *Proc. of the Intern. Offshore and Polar Eng. Conf.* 1863
[4] 2018 Safety guide *Methodological Recommendations on the Classification of Man-made Events in the Field of Industrial Safety at Hazardous Production Facilities of the Oil and Gas Industry* Approved by the Order of the Federal Service for Environmental, Technological and Nuclear Supervision of January 24, 2018 No. 29 [Electronic resource]. - Access mode: http://sudact.ru/law/prikaz-rostekhnadzora-ot-24012018-n-29-ob/rukovodstvo-po-bezopasnosti-metodicheskie-rekomendatsii/ (access date: 1.04.2018)
[5] Sukharev M G 1994 *Reserving of Main Pipeline Systems, Reliability of Gas and Oil Supply Systems* ed M G Sukharev Book 1 **3** (Moscow: Nedra) 414 p
[6] Sukharev M G and Stavrovskiy E R 1987 *Reserving of Main Pipeline Systems* (Moscow: Nedra) 167 p
[7] Sennova E V et al 1994 *Reliability of Energy Systems and their Equipment, Reliability of Heat Supply Systems* **4** (Moscow: Energoatomizdat) 480 p
[8] Sukharev M G 2010 *Methods of Forecasting* (Moscow: MaxPress) 76 p
[9] Antonov G N et al 1994 *Reliability of Energy Systems and their Equipment* ed Yu N Rudenko **1** (Moscow: Nedra) 414 p
[10] Kredentser B P 1978 *Forecasting the Reliability of Systems with Time Redundancy* (Kiev: Naukova dumka) 240p
[11] 1983 *Questions of the Mathematical Theory of Reliability* ed B V Gnedenko (Moscow: Radio and Communication) 376 p
[12] Pevzner L D 2013 *Theory of Control Systems* (St. Petersburg: Lan) 424 p
[13] Belov V V 1976 *Theory of Graphs* (Moscow: Vysshaya Shkola) 391 p
[14] Diestel R *Theory of Graphs* (Novosibirsk: Institute of Mathematics Press) 2002 336 p

IOP Conf. Series: Materials Science and Engineering **445** (2018) 012008 doi:10.1088/1757-899X/445/1/012008

Condition for the formation of mechanisms of increasing gas emission in the supply nozzle of a swirling chamber

E N Kabes[1], V V Golik[1] and A A Gladenko[2]

[1]Industrial University of Tyumen, 38 Volodarskogo st., Tyumen, 625000, Russia
[2]Omsk State Technical University, 11 Mira Prospekt, Omsk, 644050, Russia

E-mail: rechkino58@mail.ru

Abstract. In this paper, the degassing process in a swirling chamber as a function of the pressure drop, the diameter and length of the nozzle, and the mass flow velocity was considered. On the basis of the equations of the velocity of the pressure drop in the nozzle and the Rayleigh-Lamb equation describing the gas bubble growth, the dependence of the pressure change in the nozzle on time was obtained and the critical area was determined in which the formation of gas bubble nuclei occur. The condition of avalanche increase in gas emission in certain intervals of the nozzle length is obtained and its optimum value is determined. This condition makes it possible to realize the mechanism of heterogeneous and barrier-free gas evolution in a swirling chamber.

1. Introduction

One of the most important tasks of the oil and gas industry is the process of separation of multiphase (gas-liquid and liquid-gas) mixtures into separate phases, as well as physical processes which facilitate the separation of these mixtures.

2. Methodology

The analysis of the separation process in a swirling chamber is based on the assumption of the existence of the phase equilibrium in the system. However, in fact, dynamic processes are no less significant, since the gas phase formed in the liquid does not immediately separate from the liquid, due to the stretching of the process itself over time. We can expect the formation of bubbles at the stage of the introduction of the liquid phase into the swirling chamber. In this case, the formed gas bubbles in the nozzle, entering the swirling chamber, can become nuclei of heterogeneous nucleation. For a barrier-free, i.e. avalanche increase in gas emission within the swirling chamber, it is necessary to ensure a critical value of the degree of supersaturation.

The investigation of the gas bubbles formation with a rapid change in the pressure in the liquid must include the equations of conservation of mass, momentum, and energy for both liquid and gas bubbles. The equations for the formation of bubble nuclei describing their dynamics, considering the interphase energy transfer, can be obtained by following the results of the paper [1]. To consider the problem of the appearance of gas bubbles in a laminar liquid flow, the traditional approach (Lagrangian-Eulerian) to describing a two-phase medium [2] was used. It is assumed the phase equilibrium and the evolution of bubbles with the help of a model of test particles. On the basis of the equations of the pressure drop rate in the nozzle and the Rayleigh-Lamb equation describing the

IOP Conf. Series: Materials Science and Engineering **445** (2018) 012008 doi:10.1088/1757-899X/445/1/012008

growth of a gas bubble, the dependence of the pressure change in the nozzle on the time can be obtained.

The equation of the gas mass conservation inside the bubble is:

$$\frac{\partial P}{\partial t} \approx -\frac{3P}{R}\frac{dR}{dt} + \frac{\eta_a}{R}\sqrt{\frac{R_* T}{2\pi}}(|P_s - P|)\theta \ , \qquad (1)$$

where θ - dimension factor

The Rayleigh-Lamb equation:

$$R\frac{d^2R}{dt^2} + \frac{3}{2}\left(\frac{dR}{dt}\right)^2 = \frac{1}{\rho_l}\left(P - P_l - \frac{2\delta}{R} - \frac{4\mu}{R}\frac{dR}{dt}\right), \qquad (2)$$

where μ - fluid viscosity

At the initial instant of time, a homogeneous liquid which does not contain bubbles is at a set pressure and liquid temperature

$0\mathrm{t} = 0;\ P_\ell = P_{\ell 0},\ T_\ell = T_{\ell 0},\ N_b = 0,\ U = 0$

In the section of the supply nozzle, a pressure drop is set

$P_\ell = P_{\ell 0} - V_p t$, where, υ_p - pressure drop rate.

Conditions change after the formation of bubbles: $R = R_{cp}$, $P = P_s$ A qualitative analysis of the solution is shown in Figure 1.

It is clearly seen that the pressure in the selected section of the nozzle drops to values less than P_s.

In this case, the liquid appears in the metastable state, which leads to a fluctuation formation of the nuclei followed by the intensive growth of the gas nuclei. As a result of the last process, the pressure stabilizes near the value P_s. The depth of the pressure undershoot $(\Delta P)^*$ determines the efficiency of the degassing process.

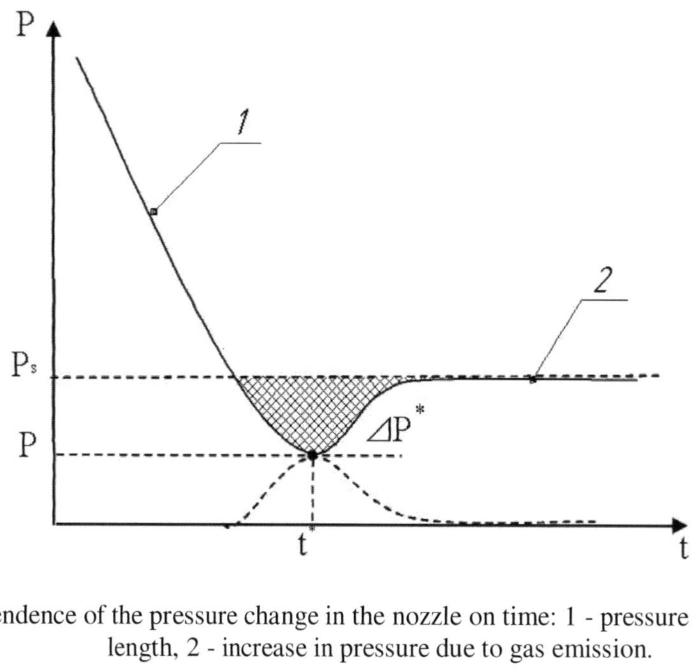

Figure 1. Dependence of the pressure change in the nozzle on time: 1 - pressure drop along the nozzle length, 2 - increase in pressure due to gas emission.

$(\Delta P)^{*}=(P_s - P)$ - The depth of the pressure drop required for the avalanche increase in gas emission. In fact, the depth of the pressure undershoot ΔP^{*} is the driving force which determines the degree of metastable state.

The results obtained make it possible to determine the limitations on the geometric dimensions of the supply nozzle. It is known that the velocity of the gas-liquid flow is related to the pressure drop by the relation:

$$U \approx \left[(1-0{,}8k)\exp\left(-0{,}06\Delta q\frac{\ell}{d}\right)\right]\tilde{\chi}\sqrt{\frac{2}{\rho_\ell}(\Delta P)}, \tag{3}$$

where, U - gas-liquid flow rate; k – function of the ratio of the gas solubility in a given liquid and the solubility coefficient of carbon dioxide in water; Δq - the saturation degree of the liquid with gas; $\frac{\ell}{d}$ - the ratio of the nozzle length to its cylindrical diameter; $\tilde{\chi}$ - nozzle flow rate; ρ_ℓ - density of degassed liquid, kg/m^3; ΔP - differential pressure at the inlet and outlet of the nozzle, Pa.

The average residence time of the flow in the nozzle $\bar{t}$ can be determined by the expression

$$\bar{t} \approx \frac{\Delta P}{\left(\frac{dP}{dt}\right)} \tag{4}$$

$\frac{dP}{dt} = \upsilon_\rho$ - the set pressure drop rate.

Then the length ℓ and the speed will be related by expression $\ell = U \cdot \bar{t}$.

Consequently, from expression (3), considering (4) and the results of calculating the pressure drop and (2), the condition for ΔP can be obtained at which the regime of avalanche increase of gas evolution in the nozzle is conducted.

$$\Delta P \approx \left(\frac{\ell}{\bar{t}}\right)^2 exp\left(-0{,}12\frac{\Delta q\ell}{d}\right)\frac{\rho_\ell}{\mu^2} > \Delta P^{*}, \tag{5}$$

In Figure 2 the dependence $\Delta P(\ell)$ is shown from which it follows that there is an optimal size of the supply nozzle for each degree of supersaturation.

$$\left(\frac{\ell^{*}}{d}\right) = (0{,}06\Delta q)^{-1}, \tag{6}$$

As follows from equation (5) for small values of the nozzle length, the determining value for the pressure drop is its quadratic dependence (curve 1) in Figure 2. With increase in the nozzle length, the second factor (curve 2) acquires an increasingly essence exponentially rapidly decreasing dependence $\Delta P(\ell)$.

As a result, a dome-shaped dependence is established, from which the existence of the minimum and maximum values of the nozzle size follows required for the realization of the heterogeneous degassing mechanism.

In Figure 2 it is shown that for a specific value ΔP^{*}, set by solutions (2) and (3), there is an interval of the geometric nozzle size for the effect realization.

$$\ell_{min} < \ell < \ell_{max}, \tag{7}$$

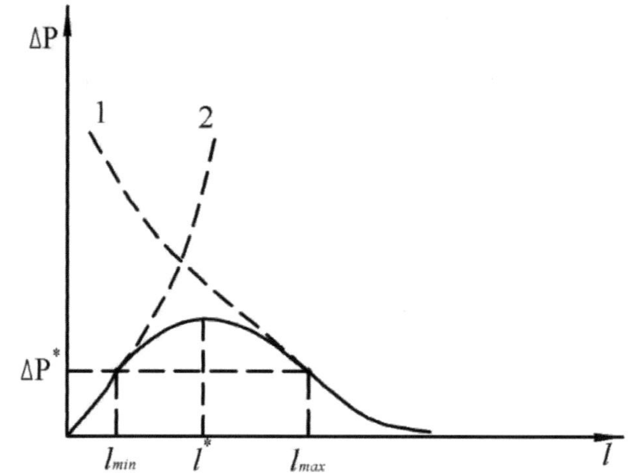

Figure 2. Graph of the dependence of the pressure drop in the nozzle as a function of its length: the curve which determines the quadratic dependence (the first factor in equation (5), 2 is the curve determined by the exponential factor in expression (5).

3 Conclusion

Thus, it was possible to identify the conditions under which the pressure drop wave propagates inside the supply nozzle below the pressure corresponding to the value at the spinodal, which determines the metastable state of the solution. As a result, in the nozzle, behind the front of the discharging wave, there will be an intensive process of gas evolution. Consequently, a solution containing gas bubbles will enter the swirling chamber, causing a hetero-phase character of the gas emission in the swirling chamber.

References

[1] Boyko V. G, Mogel H P, Sysoev V M and Chalyi A V 1991 Features of metastable states in liquid-vapor phase transitions *UFN* **161** pp 72-111

[2] Goldshtik M A 1981 *Swirling Flows* (Novosibirsk: Nauka SB RAS) 365 p

[3] Kumzerova E Yu and Schmidt A A 2002 Numerical simulation of nucleation and bubble dynamics with a rapid drop in fluid pressure 7 *ZhTF* **72** pp 36-40

[4] Merkulov A P 1969 *Swirling Effect and its Application in Engineering* (Moscow: Mashinostroyeniye) 320 p

[5] Nigmatulin R I 1987 *Dynamics of Multiphase Environments* (Moscow: Nauka) **2** 359 p

[6] Smulskiy I I 1992 *Aerodynamics and Processes in Swirling Chambers* (Novosibirsk: Nauka) 300 p

[7] Skripov V P 1972 *Metastable Liquid* (Moscow Nauka) 250 p

[8] Skripov V P 1980 *Thermophysical Properties of Liquids in a Metastable State* (Moscow: Atomizdat) 270 p

[9] Sinaiskiy E G 1990 *Separation of Two-Phase Multicomponent Mixtures in Oil and Fas Field Equipment* (Moscow: Nedra) 272 p

[10] Frenkel Ya I 1975 *Kinetic Theory of Fluid* (Leningrad: Nauka) 342 p

[11] Bakhmat G V and Kabes E N 2007 Principles of calculating the geometric parameters of a swirling chamber to provide conditions for an avalanche increase in gas emission in a gas-liquid flow *News of Higher Educational Institutions. Oil and Gas* 3 (Tyumen: TyumGNGU) pp 47-51

[12] Bakhmat G V, Pakharukov Yu V and Kabes E N 2007 *Separation of Gas-Liquid Mixtures in Swirling Devices* (Tyumen)

Evaluation of the state of pipeline system using the exergy analysis

L M Markova

Industrial University of Tyumen, 38, Volodarskogo str., Tyumen, 625000, Russia

E-mail: markova@tyuiu.ru

Abstract. Using the foreign and Russian experience of the exergy analysis of energy systems, a method is proposed for assessing the efficiency of gas main pipelines with the help of exergy characteristics. Since the exergy characteristic of the installation uniquely determines the measure of the process reversibility, this method of energy expenditure distribution can be considered objectively justified and sufficiently reflecting the energy efficiency of each unit. The exergy expenditure by the unit or exergy characteristic can be calculated by the proposed dependencies.

1. Introduction

The gas transportation system of Russia has a number of features which determine the increased risk of its operation. Multiline main gas pipelines, assembled in technological corridors, create a high concentration of power-generating pipeline capacities. Gas pipelines and corridors have intersections with each other and oil pipelines. The reliability and safety of such units are subject to special requirements. Concentration of energy objects creates a natural zone of increased risk. The increased risk of pipeline operation is objectively related to their age, as the aging of pipelines causes a reduction in the protective properties of insulating coatings, the accumulation and development of defects in pipes and welded joints as well as aging of pipe metal [1].

2. Methodology

During the operation of pipelines under the influence of workloads, temperature, atmospheric precipitation and corrosive substances, the state of the pipes and the reliability of their operation change.

Design errors and defects made during the manufacture of construction, installation and repair work manifested in the operation process in the form of failures that reduce the efficiency of the functioning of pipeline systems. When considering the causes leading to pipeline failures, it is important that a significant part of pipe failures is of a random nature and not directly related to pipe properties. They are usually caused by the presence of hidden defects, which, under the influence of internal and external factors, lead to failures.

During operation of main gas pipelines, the regularity of the change in the mechanical properties of the base metal and welded joints over time was established. During the first 5-10 years of operation of the gas pipeline, the pipes properties remain practically unchanged. The decrease in the plastic and viscosity properties of the base metal and welded joints occurs in the second time dependence interval exceeding 10-15 years [4].

Statistics on the destruction of pipelines and special studies of pipe steels and welded joints have shown the effect of loading time and operating conditions on the change in the metal properties. The research results led to the formulation of a comprehensive task of assessing the pipeline condition, taking into account the time and technological parameters of operation, as well as the initial properties of pipe steels. The diameter and working pressure of the pipelines determine their through-put capacity, which is practically proportional to the maximum operating pressure. Although some disturbance is observed due to the influence of the compressibility coefficient, this effect is insignificant at positive gas transport temperatures and pressures of 7.5 ... 12 MPa. The thickness of the gas pipe wall, as well as the diameter, determines the metal costs. Since long-term operation results in local thinning of the pipe walls caused by the action of the environment (corrosion, erosion or mechanical damage), the optimality criterion must be supplemented by another indicator. In the opinion of many experts such an indicator may be a minimum of metal specific costs, defined as the ratio of metal to the amount of transported gas [7]. Analysis of the ratio of options for gas pipelines that differ in terms of service life, conditions and parameters of operation and the strength properties of the metal can be made using the expression:

$$\Psi = \frac{g_2}{g_1} = \left(\frac{D_1}{D_2}\right)^{0,5} \frac{\sigma_{вр1}}{\sigma_{вр2}} \frac{(K_1)_2}{(K_1)_1} \frac{(K_н)_2}{(K_н)_1}, \tag{1}$$

where g – unit costs of metal; D – outer diameter of gas pipeline; K_1 – safety factor for material; $K_н$ – reliability index; indices 1 and 2 indicate the option number.

Analysis of the above mentioned expression allows us to conclude that the increasing in the diameter of the gas pipeline leads to a decrease in the specific metal costs. The same trend is observed when the temporary resistance increase in and improvement in the quality of the metal pipes. The working pressure increase is impossible without increasing in the pipe material strength. Therefore, the priority direction of reducing metal costs is the efficiency increase of the pipe steels and preserving it in the exploitation process. The direction was realized on the basis of studies of domestic and foreign authors, which resulted in the belief that the improvement of hydrocarbon transportation technology and maintenance of the unit is the most priority direction of maintaining the pipeline's working capacity and increasing its residual resource.

The efficiency of gas pipelines is a metal property to withstand avalanche brittle and extended viscous fractures. As a normative characteristic of the mentioned feature, the work of development of destruction a_p is assumed. This value characterizes the pipe material resistance to the crack development and can be determined from expression

$$a_p = \frac{\pi(1-\mu^2)}{8E} \frac{P^2 D^3 n^2}{\delta^2} \left[1 - \left(1 - e^{-3c}\right)^2\right], \tag{2}$$

where μ – Poisson's ratio, $\mu = 0,25...0,33$; P – operating pressure; D – diameter of the pipeline; n – overload factor; c – sound velocity; $\pi/(8E) = 0,0017$; E – standard modulus of steel elasticity; δ – pipe wall thickness.

The fracture development is a characteristic which reflects the state of the material in the unit, but does not take into account the system of performance indicators, in particular, the loading and technical condition of the pipeline, which depend on the productivity of the gas pipeline, determined by seasonal unevenness of gas consumption, the state of the environment and the consistency of the system: field – pre-operation unit – main pipeline – gas supply system. During the pipeline operation, various types of energy are transformed. The amount of elastic energy in the metal changes and the electrical or mechanical energy is converted into the gas flow or thermal energy. Thus, the load factors, the technical condition of the pipeline and the destruction development to a certain extent are indicators of the energy perfection of the gas pipeline. For the analysis of the pipeline condition, it is possible to propose a general thermodynamic method which makes it possible to evaluate the efficiency of technical processes in a convenient and visual way. The implementation of the method is carried out by compiling the exergy system balance, since the exergy assessment takes into account the interaction of the system in question with the environment [5, 8 and 9]. The technical value of energy depends not only on its own parameters, the form and state of the object which it possesses, but also on the parameters of the environment.

Energy research methods were proposed by Griffiths for spontaneous destruction of the pipeline. They were further refined and developed by many researchers. Since the basis of the energy research methods is the first law of thermodynamics, then considering the pipeline as an isolated system, it can be assumed that the energy of an isolated system remains constant for all changes occurring in this system. At present, there is no acceptable theory of the energy evaluation of pipeline systems which adequately describes the various states. The need for such a theory is confirmed by studies of the scale effects of the energy nature which take place on pipelines of considerable length and high throughput.

Exergy as a general indicator which determines the maximum ability to perform work allows evaluation of the energy quality (its practical availability). To establish the zero level of the mentioned indicator, it is necessary to take into account the influence of the environment on the flow of all processes during the transportation of hydrocarbon raw materials. The zero level of the overall quality index is determined by the conditions of thermodynamic equilibrium with the universal components of the environment. The system's exergy is the maximum ability of this system to perform work in such a process, which final state is determined by the condition of thermodynamic equilibrium with the environment [2, 3].

For an adapted assessment of the pipeline quality (its practical availability), it is proposed to introduce a general indicator which determines the maximum ability to perform work. To establish the zero level of the mentioned indicator, it is necessary to take into account the influence of the environment on the flow of all processes during the transport of hydrocarbon raw materials. The zero level of the overall quality index is determined by the conditions of thermodynamic equilibrium with the universal components of the environment. The system's exergy is the maximum ability of this system to perform work in such a process, which final state is determined by the condition of thermodynamic equilibrium with the environment.

Using the foreign and Russian experience of the exergy analysis of energy systems, a method is proposed for assessing the efficiency of a pipeline using exergy characteristics [10].

Loss can be produced according to their exergy characteristics. Since the exergy characteristic of the installation uniquely determines the degree of reversibility of the process, this method of energy expenditure distribution can be considered objectively justified and sufficiently reflecting the energy efficiency of each unit. The exergy expenditure by the unit or the exergy characteristic can be calculated by the formula:

$$E_X = E_i' - E_i'' + E_q' - E_q'' - W' - W'' \, , \tag{3}$$

where E_i', E_i'' - the amount of exergy introduced into the unit and carried out from it by flows of various materials,

E_q', E_q'' - the amount of exergy of the heat flow introduced into the installation and taken out from it

W', W'' - the amount of electrical or mechanical energy is equal to the exergy, respectively consumed and produced by the unit.

The energy costs for the process implementation can be calculated by the average cost of exergy consumed by the pipeline system:

$$Ц_X = \frac{Ц_j(E_i' - E'') + Ц_q(E_q' - E_q'') + Ц_W(W' - W'')}{E_X}, \tag{4}$$

where $Ц_j$ $Ц_j$ $Ц_j$ - costs of the unit of exergy, those of the transported product, heat flow and electricity respectively.

3. Conclusion

Thus, the use of the exergy analysis method for pipeline systems can be considered objectively justified and making it possible to determine and evaluate the system efficiency in order to predict the residual resource.

References

[1] 2010 *Operation of Main Gas Pipelines* ed Yu D Zemenkov (Tyumen: TSOGU) 534 p
[2] Brodyanskiy V M 1973 *Exergy Method of Thermodynamic Analysis* (Moscow: Energia) 296 p
[3] Brodyanskiy V M, Fratsher V and Mikhalek K 1988 *Exergy Method and its Applications* ed V M Brodyansky (Moscow: Energoatomizdat) 288 p
[4] Yu D Zemenkov et al 2013 Modern Management and Innovative Scientific and Technical Policy (Saint-Louis, MO: Publishing House Science & Innovation Center) 427 p
[5] Simonov V V and Veremenko A P 1998 Assessment of the state of pipeline systems using exergy method *Oil and gas* 4
[6] Glebova E V and Glebov L S 2001 *Fundamentals of Resource and Energy Saving Technologies of Hydrocarbon Raw Materials* (Moscow: State Unitary Enterprise "Oil and Gas", Russian State University of Oil and Gas n.a. I.M. Gubkin) 173p
[7] Ivantsov O M and Dvoiris A D 1980 *Low-Temperature Gas Pipelines* (Moscow: Nedra)
[8] Levin L A 1988 The optimal solution when calculating the exergy-net, taking into account the term of the energy payback *Increase in Efficiency and Optimization of Heat Energy Units. Coll. of Sc. Papers* (Saratov: SPI) pp 29-33
[9] Chukhin I M 2006 Exergy analysis of the economics of thermal power plants *Proc. on Heat Power Engineering NKTE-2006* 2 (Kazan: KazNTS RAS) pp 51-4
[10] Zagoruchenko V A et al 1980 *Thermodynamic Calculations of the Transportation and Regasification Processes of Natural Gases* (Moscow: Nedra) 320 p

IOP Conf. Series: Materials Science and Engineering **445** (2018) 012010 doi:10.1088/1757-899X/445/1/012010

Improving the safety of operation of tanks with a floating roof in the winter period

A M Kulikov[1] and S M Chekardovskiy[1]

[1] Industrial University of Tyumen, 38 Volodarskogo str., Tyumen, 625000, Russia

E-mail: kulikovam@tyuiu.ru

Abstract. Tanks with a floating roof (VST FR) allow reducing losses from evaporation of light oil fractions and petroleum products by 90-98%. However, their use in areas with high snow cover, which includes the Tyumen Oblast, is difficult because the accumulation of a large amount of snow on the roof can lead to sinking, and uneven distribution of the snow cover causes its skew and further jamming on the rails racks. Such accidents have occurred and caused enormous damage.

1. Introduction

The export of oil, natural gas and petroleum products is the main source of income of the budget of the Russian Federation. Effective use of these energy resources is impossible without the constant updating and development of technologies used in the exploration, production, transportation and storage of hydrocarbons. Over the years, steel vertical cylindrical tanks have been used to store oil and petroleum products. In our country there are more than 50,000 such structures.

The expected decline in world oil reserves, the rise in the cost of its extraction with the offshore production, the transfer of these processes to northern and other areas with complex natural and climatic conditions require the creation of new tank farms.

In our country, due to difficult climatic conditions, vertical cylindrical tanks with a fixed roof with a pontoon or without (VSTP, VST) are most widely used. Conical and spherical stationary roofs are designed and widely used. Such roofs are technologically productive, effective under uniform loads. However, the most technological is the design of a floating roof tank, both single- and double-deck.

2. Materials and methods

Tanks with a floating roof (VST FR) allow reducing losses from evaporation of light oil fractions and petroleum products by 90-98%. However, their use in areas with high snow cover, which includes the Tyumen Oblast, is difficult because the accumulation of a large amount of snow on the roof can lead to sinking, and uneven distribution of the snow cover causes its skew and further jamming on the rails racks. Such accidents have occurred and caused enormous damage.

A normative document [1] regulates the allowable height of the snow cover for tanks in operation, depending on the snow regions of the Russian Federation. The allowable height of the snow cover is determined for both uniform and uneven distribution of snow on the floating roof. The methods and recommendations for the calculation of floating roofs are given in the following effective regulatory documents: GOST 31385-2008 "Vertical cylindrical steel tanks for oil and petroleum products. General technical specifications", RD-23.020.00-KTN-079-09 "Rules for repair and reconstruction of oil storage tanks with a volume of 1000 - 50000 cubic meters", STO-SA-03-002-2009 "Rules for the design, manufacture and installation of vertical cylindrical steel tanks for oil and petroleum products."

To date, the surface of the floating roofs of tanks is cleaned manually, if a snow thickness is more than 100 mm. The process of cleaning the snow is very laborious, since a complex snow-ice mixture can form on the floating roof, which at a thickness of 10 cm on a tank of 50,000 m3 weighs about 28.7 tons. Experience of the operation of floating roof tanks has shown that the snow cover on the floating roof is distributed unevenly over the surface. This is due to the influence of wind during snowfall. Uneven distribution of snow loads leads to a heeling moment, which contributes to a failure in operability or even to flooding of the floating roof. The cases of drowning of floating roofs of tanks, which were exploited in areas with a snow load of 1.5 kPa and above, were noted. The uneven distribution of the snow load is due to the geometric parameters of the tank, the velocity and direction of the wind flow during snowfall, the height of the liquid level in the tank.

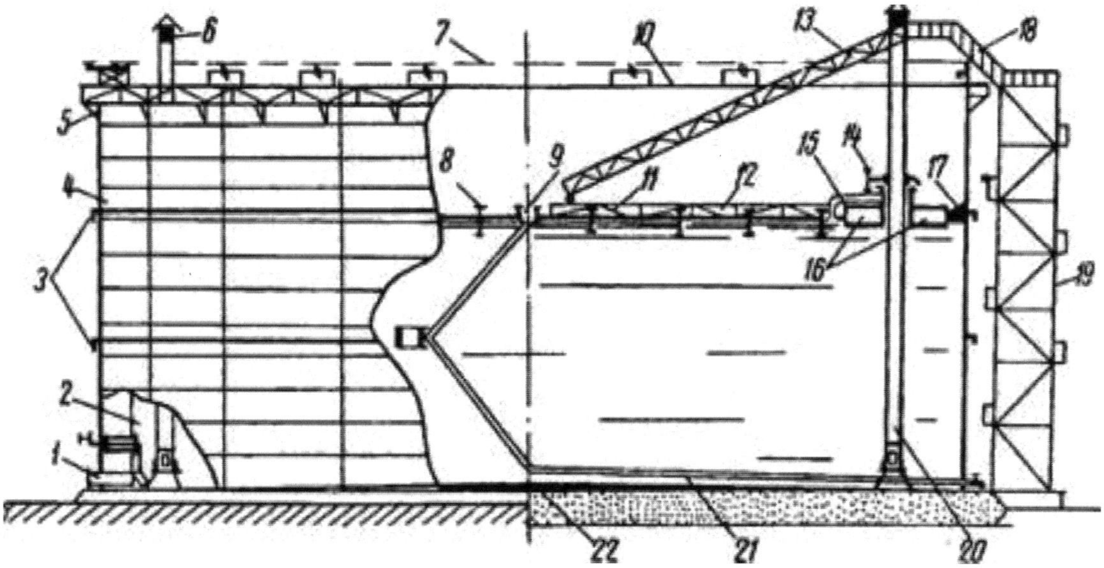

Figure 1. Structure of a VST with a floating roof.

1 - receiving and distributing branch pipe with gate valve; 2 - spare gate valve cable; 3 - stiffening rings; 4 - tank wall; 5 - annular stiffening platform; 6 - flame arrester; 7 - pipeline of foam solution; 8 - supporting pillars of floating roof; 9 - water intake of atmospheric precipitation; 10 - pipeline of irrigation of the tank wall; 11 - floating roof; 12 - supporting truss; 13 - reeling ladder; 14 - edge of foam retention; 15 - supporting truss; 16 - peripheral ring pontoon of floating roof; 17 - the seal (shutter) of floating roof; 18 - transition platform; 19 - shaft ladder; 20 - tubular floating roof guide; 21 - drainage system; 22 - tank bottom

Cleaning the roof of even one tank is very labor-intensive. At present, the cleaning procedure is carried out as follows: the tank to be cleaned is filled with oil to the maximum level to raise the roof, then a team of 4-5 people manually shovels the roof for four days. After cleaning is complete, the procedure is repeated for other tanks at the farm. The area of the floating roof of a VST FR of 20,000 m^3 volume is more than 1600 square meters, hence such a long time. For example, in Surgut UMN, VST roof cleaning is carried out about five times during the winter period.

A possible solution to this problem may connected to the melting of snow due to the heat of the stored product, namely oil. Here, it is necessary to calculate the balance of energy required for melting snow on the roof, and the heat flow from the oil. In this paper, the process of melting snow is considered at the time of its fall, with water draining into the drain pipe of the drainage system 21, Figure 1

The energy required to melt snow on the metal surface of the roof is composed of the following quantities:

1) Energy input for heating snow from air temperature to 0°C

$$Q_{sn}=G*C_{sn}*(t_{melting} - t_{air}) \quad (1)$$

where G – snowfall rate, kg/hour, we choose from the "Construction climatology" and the ratings of precipitation in the region;

C_{sn} – heat capacity of snow, C_{sn} = 2100 $\frac{kJ}{kg*k}$;

t_a – average air temperature over the period under study.

2) Energy input for melting ice:

$$Q_{melt}=G*\lambda \quad (2)$$

where λ – specific heat of melting ice, λ=335 $\frac{kJ}{kg*k}$;

3) Energy input for heating water (heating water after melting snow is necessary for draining it into the drainage system of the tank):

$$Q_{hwr} = C_{wr}*G\,(t_{water} - t_{melting}) \quad (3)$$

where C_{wr} –heat capacity of water, C_{wr} =4200 $\frac{kJ}{kg*k}$

In this case, it is necessary to consider the entrainment of thermal energy Q from the surface of a floating roof by wind.

$$Q = \alpha*S_{roof}*(t_{roof} - t_{air}) \quad (4)$$

where α – coefficient of heat transfer from the roof;

S_{roof} – roof area;

$$\alpha = Nu_a*\lambda_a/l \quad (5)$$

where λ_a – coefficient of thermal conductivity of air.

l – linear dimension. This parameter depends on the value of the roof surface unoccupied with snow which prevents heat loss and is a natural heat insulator;

$$Nu_a = 0{,}032*Re_a^{0,8} \quad (6)$$

$$Re_a=\frac{w*l}{\nu_a} \quad (7)$$

where w – wind velocity above the roof surface;

ν_a – kinematic viscosity of air.

Calculations show that at an air temperature of -20^0 C, for the floating roof of RVSPK-20000, the heat flow density required for operational snow heating is about 140 W/m^2.

In this paper, we will consider heat transfer from oil through a two-deck roof, since its calculation represents maximum interest and complexity.

The main types of floating roofs currently in use are shown in Figure 2:

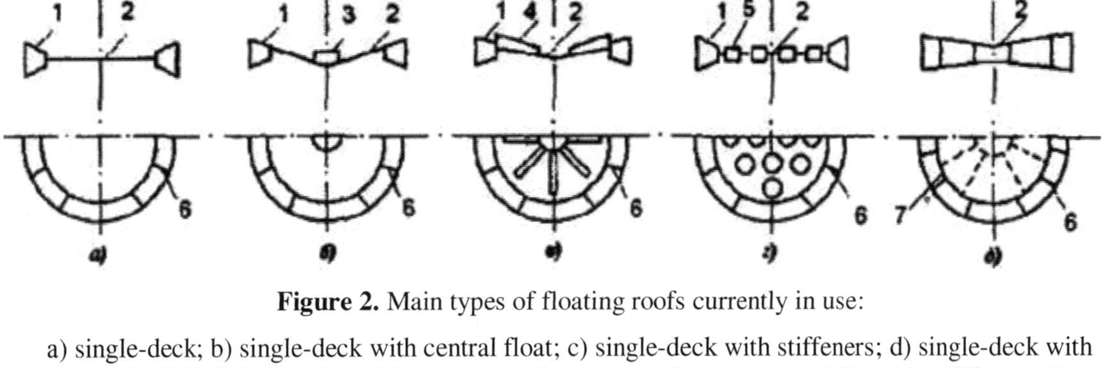

Figure 2. Main types of floating roofs currently in use:

a) single-deck; b) single-deck with central float; c) single-deck with stiffeners; d) single-deck with floats; e) double-deck; 1 - closed ring pontoon; 2 - sheet roofing; 3 - central float; 4 - stiffeners; 5 - float; 6 - radial bulkhead; 7 - annular bulkhead

IOP Conf. Series: Materials Science and Engineering **445** (2018) 012010 doi:10.1088/1757-899X/445/1/012010

The double-deck floating roof of the oil tank includes concentrically arranged roof segments containing interconnected compartments, the compartments being arranged such that the smaller of the tangential surfaces of the compartment of the first segment adjoins a portion of its surface to a part of the surface of the larger tangential surface of the compartment of the second segment. The roof comprises a radial frame belt that is connected to one pair of compartments of neighboring roof segments adjacent to each other by parts of their tangential surfaces, further joining them together; wherein the radial frame belt has such a length that it essentially connects the central part of the roof and the side sheet.

In a confined space, free convection develops differently, since the motion near surfaces having different temperatures is interrelated. In narrow gaps, flat and annular channels, layers of different shapes, the heat flow density q is calculated from the formulas of stationary thermal conductivity in a flat wall, introducing the concept of the equivalent thermal conductivity coefficient:

$$q = \lambda_{equiv} * \Delta t / l \tag{8}$$

where λ_{equiv} – equivalent thermal conductivity coefficient;

l – distance between the top and bottom decks of a floating roof;

Δt – temperature difference between the top and bottom decks of a floating roof.

In our case, we set the boundary temperatures of the top and bottom surfaces of the roof, namely: the top deck - +5^0 C, the bottom deck - the temperature of the oil. The choice of these temperatures is due to the following factors: the need to melt the snow and maintain the water temperature above zero for unimpeded drain into the drainage system along the top deck surface. The temperature of the bottom deck is equal to the temperature of the oil, since heat losses during the oil-wall heat exchange can be neglected.

The equivalent thermal conductivity coefficient is determined by the formula:

$$\lambda_{equiv} = \varepsilon_c * \lambda_a \tag{9}$$

where λ_a – coefficient of thermal conductivity of air in a closed compartment of a floating roof;

ε_c – convection coefficient, a correction that takes into account an increase in the heat flow due to free convection in the closed space of the floating roof compartments.

When calculating the convection coefficient ε_c, first, one should consider a change in the physical properties of air, such as the kinematic viscosity ν_a, the coefficient of temperature expansion β, the Prandtl Pr_a criterion, depending on its temperature, as they vary within fairly wide limits and strongly affect the results.

Air properties are determined by the average temperature of the closed compartment:

$$T = \frac{T_1 + T_2}{2} \tag{10}$$

where T_1 and T_2 – top and bottom deck temperatures.

The convection coefficient is calculated depending on the product of the Grashof and Prandtl criteria for air in the closed space of a double-deck roof:

- for the product of $[Gr_a*Pr_a]$ from 10^3 to 10^6 the convection coefficient ε_c is calculated by the formula:

$$\varepsilon_c = 0.105*[Gr_a*Pr_a]^{0.3} \tag{11}$$

- for the product of $[Gr_a*Pr_a]$ greater than 10^6 the convection coefficient ε_c is calculated by the formula:

$$\varepsilon_c = 0.4*[Gr_a*Pr_a]^{0.2} \tag{12}$$

Preliminary calculations show that if the temperature of the oil is maintained at +35^0 C, the heat flow from it is sufficient to exceed the energy input for snow heating during snowfall.

3. Conclusion

Calculations show that at an air temperature of -20^0 C, for the floating roof of RVSPK-20000, the heat flow density required for operational snow heating is about 140 W/m^2.

Preliminary calculations show that if the temperature of the oil is maintained at +35^0 C, the heat flow from it is sufficient to exceed the energy input for snow heating during snowfall.

References

[1] OR-23.020.00-KTN-279-09 *Special Regulations for the Operation of Single-Deck and Double-Deck Floating Roofs of Tanks RVSPK, ZHBRPK in Winter* (Moscow: Transneft) 2009

[2] Operation of Main and Technological Oil and Gas Pipelines. Distribution and Accounting 2014 (Tyumen: TyumGNGU) 370 p

[3] Zemenkova M, Shalay V, Zemenkov Y and Kurushina E 2016 Improving the efficiency of administrative decision-making when monitoring reliability and safety of oil and gas equipment *MATEC TPACEE 2016*

[4] Vlasenko V S, Slesarenko V V, Gulkov A N and Lapshin V D 2015 Preparation of arctic oil and gas condensate deposit formation fluid for storage and transport in the form of hydrocarbon hydrate-containing dispersed system *Proc. of the Intern. Offshore and Polar Eng. Conf.* 1863

[5] Zemenkova M, Zemenkov Y, Gladenko A and Podorozhnikov S 2016 Estimation of emissions during monitoring of pipelines in the dynamic mode of operation *5th Intern. Sc. Conf. Integration, Partnership and Innovation in Construction Science and Education* **86**

[6] Gorelik J B, Shabarov A B and Sysoyev Yu S 2016 The dynamics of frozen ground melting in the influence zone of two wells *Earth's Cryosphere* 12 (1) pp 59

Modernization of the individual device for temperature stabilization of the soil

K S Voronin, D A Cherentsov, A U Yakupov and Yu D Zemenkov

Industrial University of Tyumen, 38 Volodarskogo str., Tyumen, 625000, Russia

E-mail: ya.yakupov-azamat@yandex.ru

Abstract. The authors of the article developed a finite-element model of a soil with a thermal stabilizer that takes into account the thermal effect from the environment, as well as the temperature of atmospheric air, wind speed, magnitude of the heat flux from solar radiation, albedo of the active surface, magnitude of the heat flux in the infrared, thermal conductivity of the snow cover, thermal conductivity and heat capacity of the soil, which makes it possible to calculate the haloes of freezing the soil. The results of a numerical experiment conducted using the COMSOL Multiphysics software complex showed that using the developed design of the thermal stabilizer allows increasing the zone of soil freezing around the bottom of the cooling device.

1. Introduction

Active reclamation of the regions of the Russian Federation covered with permafrost soils (PFS) requires the application of protection means for rocks that have been in a frozen state for dozens of years. Such rocks have specific physical, mechanical, engineering-geological, filtration properties that are taken into account when designing the bases, foundations for the construction of facilities on the territory of PFS. The permafrost rocks have an upper active layer, which during the year changes its temperature, and the lower one, which has a negative temperature. Because of the abundance of ice, rock characteristics change with temperature. An inadmissible condition for the operation of bases and foundations are deformations that may exceed the limit values for the melting of PFS.

The regulatory document for the construction of buildings and structures on the territory of PFS is the set of rules "Foundations of Buildings and Structures" [5], according to which the projected facilities are built following one of two principles. The first principle involves the use of permafrost soils as bases while maintaining them in the frozen state throughout the life of the facilities. The second principle permits the use of permafrost soils as bases in the thawed state. According to the first principle, constructions are erected with the economic justification and its expediency, the second principle is used when the deformations do not exceed the maximum permissible values for the erected structure.

2. Materials and methods

The applied measures to protect the PFS from thawing are divided into two types [6]: active and passive.

Active measures are ventilated basements, which are used for the bases of buildings and structures. Advantages: possibility to use in combination with other cooling means. Disadvantages: it is not the main method of maintaining negative soil temperatures.

The tubular system for freezing the base soils is divided into a horizontal "HNT" and a vertical "VNT". The operational principle is based on the circulation of refrigerant through the tubes and the heat transfer from the evaporative part to the condenser part, where heat is released into the atmosphere. The HNT system. Advantages: the ability to maintain the temperature in hard-to-reach places. Disadvantages: metal-consuming method, cools the surface of the soil. The VNT system. Vertical cooling pipe system "VNT". Advantages: it freezes the foundation soil faster at the initial stages of construction and operation, and fuses the soil within 12-15 meters depth. Disadvantages: it is used mainly to maintain the temperature of tanks, roads and railways, buildings up to 120 meters wide.

Individual seasonal cooling devices "SCD" - thermal stabilizers. Devices have an individual single-tube design with an all-metal casing, charged with a refrigerant. By design features they are divided into vertical, inclined and slightly inclined. According to the type of refrigerant used: liquid, vapour-liquid, gas and gas-liquid. The vapour-liquid ones have found wide application due to good heat-transferring ability. According to the principle of operation, thermal stabilizers of seasonal and year-round action are distinguished. The principle of the thermal stabilizer is based on the circulation of the refrigerant inside the tube. The lower evaporative part of the SCD is underground. Taking the heat of the soil, the refrigerant inside the structure boils, the vapour of the boiling refrigerant rises, carrying with it heat energy. In the external condenser part, the refrigerant vapour condenses, transferring heat to the outside air. Condensed refrigerant flows down closing the continuous circulation. The refrigerant circulation is only possible if there is a temperature difference in the evaporator and condenser parts. In summer, the temperature of the outside air is much higher than the temperature in the evaporator part and the effect of the heat stabilizer is suspended. Advantages: universality, the method is applicable to any type of object that needs to maintain negative soil temperatures. Disadvantages: the dependence of the duration of the device's active operation on external factors - the ambient temperature. To eliminate the disadvantage, it is necessary to install a greater number of heat stabilizers, to prevent thawing of the soil and deformations of the structures.

In the all-year-round SCD, the Peltier effect is realized. The Peltier effect is a thermoelectric phenomenon, which is accompanied by the release or absorption of heat as electric current passes through the contact point of two different conductors. The amount of heat produced depends on the properties of the conductors, the direction of the current flow and its force. When installing the module realizing the Peltier effect in summer, the necessary temperature difference will be created for the continuous operation of the heat stabilizer. Advantages: the lack of dependence of the thermal stabilizer on weather conditions. Disadvantages: the need to provide a source of electricity for each thermal stabilizer.

Passive protection refers to the thermal insulation of the object with the contacting soil. For pipelines laid underground, additional means can be the use of a factory circular heat insulation pipeline, which significantly reduces the thermal load on permafrost soils, reducing halos of thawing. Also, additional heat-shielding screens and replacement of icy subsidence grounds in the bases of projected objects are used, which will reduce the drawdown and will not exceed the permissible values of deformations.

At present, the most demanded technical solution capable of compensating for man-made impacts on the bearing capacity of permafrost soils is the use of soil thermal stabilizers. After analysing the long-term Russian and foreign experience of using soil thermal stabilizers, several of them can be distinguished.

Figure 1. Transversely finned thermal stabilizer.

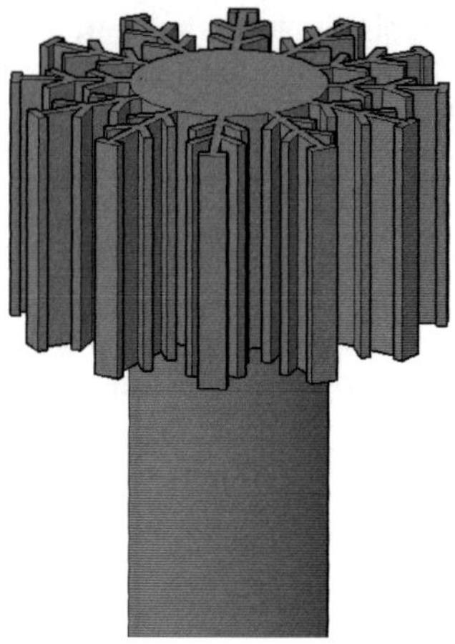

Figure 2. Longitudinally finned thermal stabilizer.

A transversely finned thermal stabilizer, shown in Figure 1, has a clear advantage in production, transportation and operation due to the simple construction of the condenser part. The main way to increase the heat transfer of the condenser part is to increase the finning area. If applying this method, metal costs increase, and with an increase in the finning coefficient, the relative depth of the intercostal gap increases, at which the root zone is gradually eliminated from heat transfer, which leads to a slowdown in heat transfer.

Longitudinally finned thermal stabilizer (Figure 2) has an increased capacity of freezing and has a greater heat dissipation capacity due to the presence of additional ribs. The design is not widely used in the industry due to the technologically complex manufacturing process and the high costs of metal production.

3. Solution procedures

The authors proposed the construction of the condenser part of the thermal stabilizer in Figure 3, which allows eliminating the identified shortcomings. It is proposed to weld tubes of a smaller diameter along the perimeter of the cylindrical surface in parallel to the base direction, which resemble the needles of a hedgehog. The ribs in the transverse plane are located at an angle to each other and as the length of the rib increases the intercostal distance between them increases either. The design features of the developed thermal stabilizer exclude the formation of stagnant zones with an increase in the finning area.

To test the advantages of the developed design of the thermal stabilizer, we will perform the numerical experiment. For this purpose, the design scheme in Figure 4 is: soil model that takes into account thermal effects from the environment, air temperature, wind speed, heat flux from solar radiation, albedo of the active surface, heat flux in the infrared part of spectrum, thickness and thermal conductivity of the snow cover, thermal conductivity and heat capacity of the soil.

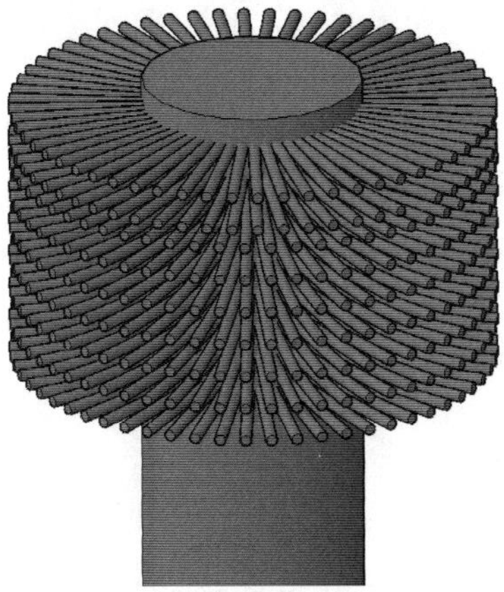

Figure 3. "Hedgehog".

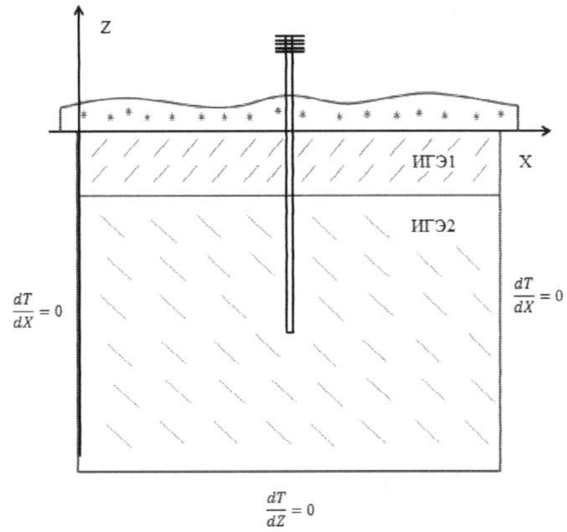

Figure 4. Calculation scheme.

Modelling of the temperature mode is performed in the Comsol Multiphysics software package. The determination of the temperature field in the computational domain requires the solution of the differential heat equation [6]:

$$\frac{\partial T}{\partial \tau} = Q\nabla^2 T + \frac{q_v}{\rho c} \quad (1)$$

Taking into account the factors considered in the developed model, we obtain:

$$p_{sk}\left(c_{p,sk} + W\left(c_{p,w} + (c_{p,l} - c_{p,w})\theta(T)\right) - L_{melt}W\frac{d\theta}{dT}\right)\frac{dT}{d\tau} = \overline{\nabla}\cdot\left(\lambda\overline{\nabla}(T)\right) \quad (2)$$

where ρ_{sk} is the density of the skeleton of the soil;
$c_{p,sk}$ is the heat capacity of the soil skeleton at constant pressure;
$c_{p,w}$ is the heat capacity of water at constant pressure;
$c_{p,i}$ is the heat capacity of ice at constant pressure;
L_{melt}- heat of the ice-water phase transition;
λ is the thermal conductivity of the soil;
θ is the ratio of the mass of ice to the mass of all water at a given temperature;
W is the mass moisture content of the soil;
T is the temperature at the point under consideration.

To solve the equation, the boundary conditions "zero heat flow" are set on the lower and lateral surfaces of the modelled soil massif and on the soil surface, according to [3]:

$$Q = Q_c - Q_s + Q_r = $$
$$= a_i(T_{pov} - T_{v,i}) - (1 - A)q + \varepsilon_i\sigma_0(T_{pov}^4 - T_{v,i}^4) + \varepsilon_i\sigma_0\eta T_{v,i}^4 \quad (3)$$

where Q is the heat flux from the ground to the atmosphere;
Q_c is the convective heat flux;
Q_s - heat flux from solar radiation;
Q_r is the radiative heat flux in the IR region of the spectrum;
α is the convective heat transfer coefficient of the underlying surface of the soil;
q - direct and diffuse heat flux of solar radiation to the horizontal surface;
ε - the degree of blackness of the underlying surface of the soil;
σ_0 is the Stefan-Boltzmann constant;
η - coefficient of atmospheric radiation;
i - the index refers to the ith month of the year.

The boundary conditions at the boundary of the SCD-atmosphere:

$$Q_v = \alpha_k(F_r E_r \mu + F_{tr})(T_0 - T_g) \tag{4}$$

where α_k - convective heat transfer coefficient;
F_r and F_{tr} - the surface area of the ribs and intercostal space;
E_r - edge efficiency coefficient;
μ is the correction factor for changing the thickness of the cross section of the rib.

The boundary conditions at the boundary of the SCD are:

$$Q_g = \alpha_v \frac{F_c}{F_e}(T - T_v) \tag{5}$$

where F_c and F_e - the area of the condenser and evaporator;
α_v_в is the heat transfer coefficient of the above-ground condenser part of the SCD [7].

4. Results of simulation

Figures 5-7 show halos of soil freezing.

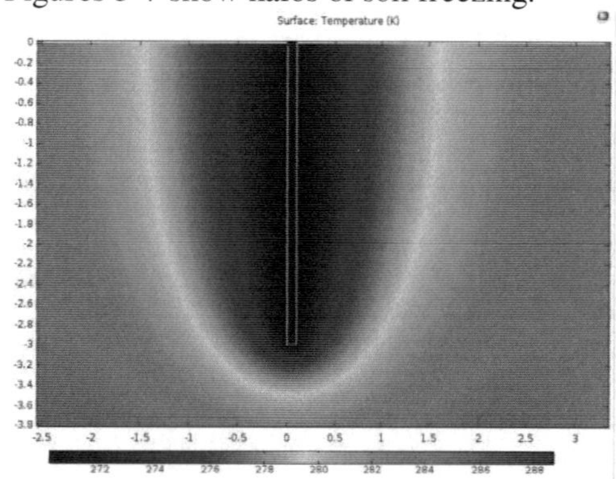

Figure 5. Transversely finned thermal stabilizer.

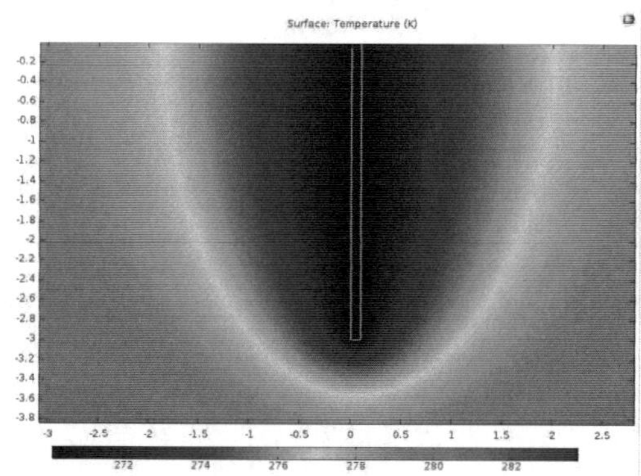

Figure 6. Longitudinally finned thermal stabilizer.

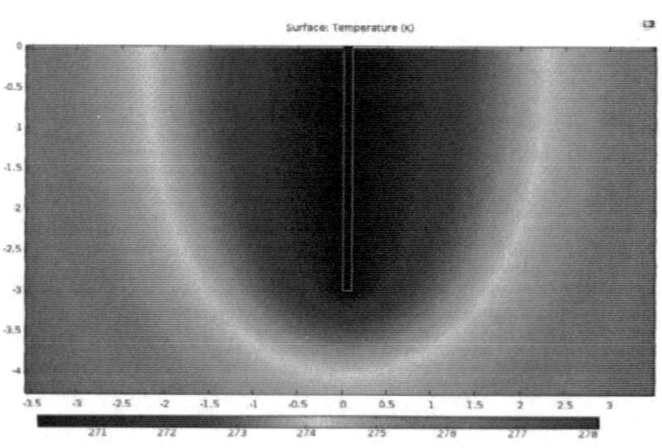

Figure 7. "Hedgehog".

In the constructions considered, Figures 1-3, the distribution of temperatures in the soil and the radius of the freezing zone around the SCD have different values. Based on the results of the numerical experiment, the halo of the freezing of the soil of the transversely finned thermal stabilizer in the radius has a maximum value of 1.5 meters, in the longitudinally finned 1.9 meters; the developed thermal stabilizer construction has 2.2 meters. The change in the shape of the ribs of the capacitor part of the SDA led to an increase in the heat transfer rate of the "heat stabilizer-atmosphere," which in turn led to an increase in the freezing halo of the soil.

Table 1. Values of the temperature distribution of the freezing halo of a transversely finned SCD.

axis Y	axis X: 0.25	0.5	0.75	1	1.25	1.5	1.75	2	2.25	2.5
-0.4	270.56	272.91	274.79	276.30	277.95	279.09	280.16	281.22	281.86	282.35
-0.8	271.49	272.9	274.7	276.44	277.92	279.27	280.35	281.26	282.00	282.46
-1.2	271.64	273.33	274.99	276.71	278.32	279.43	280.58	281.54	282.19	282.63
-1.6	271.54	273.31	275.28	276.93	278.79	279.76	280.89	281.79	282.42	282.77
-2	271.61	273.85	275.45	277.45	279.02	280.29	281.13	282.04	282.63	283.07
-2.4	271.71	274.06	275.84	278.22	279.57	280.75	281.83	282.57	283.24	283.05
-2.8	272.77	275.24	277.32	279.17	280.61	281.63	282.69	283.23	283.65	283.91
-3.2	276.33	278.06	279.33	280.67	281.79	282.76	283.46	284.02	284.26	284.40
-3.6	280.51	281.07	281.93	282.55	283.29	283.90	284.43	284.75	284.93	284.96
-4	283.44	283.51	283.98	284.34	284.78	285.08	285.40	285.55	285.57	285.47

Table 2. Values of temperature distribution of the freezing halo of the developed design of the SCD.

axis Y	axis X: 0.25	0.5	0.75	1	1.25	1.5	1.75	2	2.25	2.5
-0.4	270.56	270.92	271.41	271.89	272.36	272.82	273.25	273.71	274.12	274.53
-0.8	270.56	270.93	271.46	271.93	272.42	272.83	273.30	273.73	274.17	274.55
-1.2	270.59	270.97	271.50	271.97	272.44	272.94	273.42	273.80	274.21	274.62
-1.6	270.60	271.01	271.56	272.06	272.59	273.04	273.48	273.90	274.32	274.74
-2	270.64	271.13	271.71	272.24	272.75	273.20	273.67	274.06	274.46	274.86
-2.4	270.73	271.28	271.92	272.45	272.99	273.42	273.90	274.25	274.63	275.02
-2.8	270.97	271.66	272.32	272.85	273.32	273.73	274.16	274.51	274.89	275.21
-3.2	272.05	272.51	273.00	273.40	273.82	274.13	274.50	274.84	275.13	275.45
-3.6	273.26	273.44	273.72	273.98	274.28	274.56	274.84	275.14	275.41	275.69
-4	274.11	274.21	274.37	274.56	274.77	274.99	275.24	275.47	275.69	275.93

To find the average value of the temperature, a double integral was used in the region of the ground freezing halve, the obtained values of the temperatures of the freezing halo of the considered designs of thermal stabilizers were approximated using a polynomial. We took the double integral from the resulting function divided by the halo area and obtained the average temperature. To solve the integral, we used the mathematical simulation program Matlab.

$$\bar{T} = \frac{1}{S}\iint_S f(x,y)dA \qquad (6)$$

$$T = \int_0^{2.5}\left(\int_{-4}^{0}(267.6 + 7.67\cdot x - 4.627\cdot y + 4.235\cdot x^2 + 7.061\cdot x\cdot y -\right.$$

$$-1.637\cdot y^2 - 4.305\cdot x^3 - 0.468\cdot x^2\cdot y + 6.813\cdot x * y^2 + 1.487\cdot y^3 +$$

$$+ 1.463\cdot x^4 - 0.1925\cdot x^3\cdot y - 0.8448\cdot x^2\cdot y^2 + 1.934\cdot x\cdot y^3 +$$

$$+ 0.8976\cdot y^4 - 0.1956\cdot x^5 - 0.001187\cdot x^4\cdot x - 0.05467\cdot x^3\cdot y^2 -$$

$$-0.2146 \cdot x^2 \cdot y^3 + 0.1266 \cdot x \cdot y^4 + 0.1079 \cdot y^5)dy)\,dx$$

The calculation is made for SCD with different types of fins of the condenser part. Estimated area of the ground frost halo 5x4 m. The difference in the average temperatures of the crosswise finned and designed thermal stabilizer is 6.63 degrees.

The results of a numerical experiment conducted using the COMSOL Multiphysics software complex showed that using the developed design of the thermal stabilizer allows increasing the zone of soil freezing zone around the bottom of the cooling device.

5. Conclusions

1. The authors of the article developed a finite-element model of a soil with a thermal stabilizer that takes into account the thermal effect from the environment, as well as the temperature of atmospheric air, wind speed, the magnitude of the heat flux from solar radiation, the albedo of the active surface, the magnitude of the heat flux in the infrared, the thermal conductivity of the snow cover, the thermal conductivity and heat capacity of the soil, which makes it possible to calculate the haloes of freezing the soil.
2. It has been revealed that the average temperature of the ground freezing halo of the developed structure is 6.63 °C lower than that of the cross-finned thermal stabilizer.
3. Modernization of the condenser part of an individual seasonal device resulted in a reduction of metal costs for its production, with an increase in the heat transfer intensity, which will have a positive impact on their payback.

References

[1] Bajasan R M, Golibin S I, Lobanov A D and Bajasan T V 2012 Steam-liquid heat stabilizers of various types and purposes, their design and thermo-technical features *J. "Pipeline-Water Transport: Theory and Practice"* (Moscow: VNIIST) 4 (32)
[2] Voronin K S and Yakupov A U 2016 The urgency of the use of composite materials for the storage of oil and petroleum products in conditions of low temperatures *Oil and Gas Terminal. Mat. of the International Scientific and Technical Conference "Transport and Storage of Hydrocarbon Raw Materials"* 11 pp 36-9
[3] Markov E V, Pulnikov S A and Kazakova N V Features of the specification of boundary conditions of the third kind on the surface of multi-ice-free soil during modeling of its thermal interaction with the "hot" pipeline in the program COMSOL MULTIPHYSICS 4.3B *Fundamental Research* 11 (2) p 287-91
[4] Markov E V, Pulnikov S A and Sysoev Yu S 2015 Analysis of typical methods for calculating the depth of thawing of permafrost soils under the pipelines *Modern Problems of Science and Education* 1 p 142
[5] SP 22.13330.2011 *Soil Bases of Buildings and Structures*
[6] Roizen L I 1997 *Thermal Calculation of Finned Surfaces* ed. V G Fastovsky (Moscow: Energia)
[7] Gavra T G et al 1982 *Thermal and Hydraulic Calculation of Heat Exchangers for Compressor Units* ed A I Kirillov (Leningrad: LPI)
[8] CAN/CSA-S500-14 *National Standard of Canada. Thermosyphon Foundations for Buildings in Permafrost Regions*
[9] Denhartog S L 1988 A Thermosyphon for horizontal applications *Cold Regions Science and Technology* 15 319–21
[10] Holubec I, Jardine J and B Watt 2008 Flat loop evaporator thermosyphon foundations: design, construction, and performance in the Canadian permafrost regions *Proc. of the 9th Intern. Conf. on Permafrost* ed D L Kane and K M Hinkel (Fairbanks: Institute of Northern Engineering, University of Alaska)735–40